ÜBERREICHT VON DER

DEUTSCHEN

FORSCHUNGSGEMEINSCHAFT

BONN - BAD GODESBERG

REMEMBERING EDITH ALICE MÜLLER

ASTROPHYSICS AND SPACE SCIENCE LIBRARY

VOLUME 222

REMEMBERING
EDITH ALICE MÜLLER

Edited by

I. APPENZELLER
Landessternwarte, Heidelberg, Germany

Y. CHMIELEWSKI
Observatoire de Genève, Sauverny, Switzerland

J.-C. PECKER
Collège de France, Paris, France

R. DE LA REZA
Observatório Nacional, Rio de Janeiro, Brazil

G. TAMMANN
Universität Basel, Switzerland

and

P. WAYMAN
University of Dublin, Ireland

KLUWER ACADEMIC PUBLISHERS
DORDRECHT / BOSTON / LONDON

Library of Congress Cataloging-in-Publication Data

Remembering Edith Alice Müller / edited by I. Appenzeller ... [et
al.].
       p.   cm. -- (Astrophysics and space science library ; v. 222)
  Includes bibliographical references and index.
  ISBN 0-7923-4789-7 (hardbound : alk. paper)
    1. Müller, Edith Alice. 2. Women astronomers--Switzerland-
  -Biography.   I. Müller, Edith Alice. 'II. Appenzeller, I. (Immo),
  1940-   . III. Series.
  QB36.M78R46 1998
  520'.92--dc21
  [B]   97-48764

ISBN 0-7923-4789-7

Published by Kluwer Academic Publishers,
P.O. Box 17, 3300 AA Dordrecht, The Netherlands.

Sold and distributed in the U.S.A. and Canada
by Kluwer Academic Publishers,
101 Philip Drive, Norwell, MA 02061, U.S.A.

In all other countries, sold and distributed
by Kluwer Academic Publishers,
P.O. Box 322, 3300 AH Dordrecht, The Netherlands.

Printed on acid-free paper

Printed in the Netherlands

Edith Müller, in her office at Sauverny, with (from left to right) : Felix Llorente de Andrés, Yves Chmielewski, Ramiro de la Reza, discussing the cosmic abundances of clements...

A page out of Edith Müller's doctoral dissertation. Note the analogy with Max Escher's famous works.

TABLE OF CONTENTS

4

INTRODUCTION

FOREWORD.

Jean-Claude Pecker (Paris, France)

The idea of grouping a few texts devoted to the evocation of the life of Edith Müller came from a pleasant walk, in the sunny fall of 1996, along the beaches of the Rio de la Plata, in front of Montevideo, with Ramiro de la Reza. Edith had been, for so many of us in the world, not only an excellent colleague, but a good friend, that it became immediately obvious that the idea was quite appropriate, certainly still more for her, who had friends all over the world, than for many others. Soon, this project become more precise, with the enlargement of our little nucleus to Yves Chmielewski, her former student in Geneva, like Ramiro, Gustav Tammann, her almost neighbour in Basel, Immo Appenzeller, her... "grand-grand ....-grand son", in his capacity of General Secretary of the IAU, and finally Patrick Wayman, who succeeded Edith as General Secretary and who accepted to improve the English of these very diverse texts....

We are indebted to many friends and colleagues for their contributions, sometimes short evocations of a privileged moment, sometimes more precise and long accounts of a scientific encounter over the Sun, or over the chemical elements.... It was not the purpose of this book to give a complete account of all the scientific contributions of Edith to solar physics, nor was it intended to give a detailed view of her many administrative duties. We wanted mostly to share with the readers the remembrance we all had of a very fine person, full of life, of "joie-de-vivre", full of ideas, and devoted to her family, to her friends, to her responsibilities.

I want here not only to thank all the contributors, but also to express our appreciation of the help the IAU has given us, and to thank also Kluwer Publishers and their staff for the care they gave to the publication of this book. I personally want to thank also Ruth Schatzman who has translated the texts written in Russian, Dominique Bidois for her invaluable material help in giving a shape to that ensemble, Jean-Pierre Martin who helped us in formatting the photographs, the Observatory of Geneva, the Collège de France for various services, and of course, last but not least, my five coeditors and old-time friends Immo, Yves, Ramiro, Gustav and Patrick...

———————— Yves Chmielewski (Geneva, Switzerland) ————————
EDITH ALICE MÜLLER (1918–1995)

Short Biography

Edith Müller, a Swiss citizen, was born on February 5, 1918 in Madrid (Spain) where she attended the German School from 1924 to 1936. After having obtained her Maturity diploma in June 1936, she went to the University of Zürich, Switzerland, where she first obtained in 1942 a diploma for higher teaching (math.-phys.) In 1943, she presented, still at the University of Zürich, a quite original Ph. D. thesis on the "Application of Group Theory and Structural Analysis to the Moorish adornments of the Alhambra in Granada". From 1946 to 1951, she worked as an assistant at the Swiss Federal Observatory in Zürich.

In 1951, she was an invited astronomer at Cambridge University Observatory (Great Britain). The years 1952 to 1954 were spent as a research assistant at the Michigan University Observatory in Ann Arbor (USA). She went back to Switzerland for a short while, spending the period 1954–55 at the Basel University where she occupied a position as research and teaching associate. She returned to the Michigan University Observatory where she took over a research and teaching associateship from 1955 to 1962. It was during this very happy, active and fruitful period that she participated in the huge piece of work which was reported in the famous G.M.A. (Goldberg-Müller-Aller, 1960) paper on the abundances of the elements in the solar atmosphere, one of the most cited in the astrophysical literature. Another very important research undertaken during this period, together with J.P. Mutschlecner, concerned the effects of deviations from the local thermodynamic equilibrium on the solar abundances (Müller and Mutschlecner, 1964).

The year 1962 saw her final return to Switzerland, where she joined the young research team that had just started to revive the University of Geneva

Observatory. She was appointed Associate Professor in Geneva; and at the same time, from 1962 to 1965, held a position of Associate Professor at the University of Neuchatel Observatory. In 1972, she was promoted a full Professor at the University of Geneva, a position which she held till her official retirement in 1983. During all these years, she taught courses on the Theory of Stellar Atmospheres, Solar Physics and Astronomical Spectroscopy. She was also often called as Invited Professor by foreign universities such as Kiel (Germany), Liège (Belgium), Granada (Spain), Sofia (Bulgaria), Istanbul and Izmir (Turkey), Mexico, and, for a full year, by Utrecht University (The Netherlands). She also gave lectures in advanced astronomy summer courses organized by the IAU, UNESCO or NATO.

 In Geneva she pursued her research on the physical structure and chemical composition of the Solar Atmosphere. Her interest was mainly the study of the mean quiet Sun, with a view towards extending the methods used in the solar case to the study of the atmospheres of the solar-type stars. Occasionally she also studied some aspects of the physics of the active Sun. Her research was based on observations which she regularly carried out at the very high resolution and quality solar telescope and spectrograph of the Kitt Peak National Observatory in Arizona (USA), as well as that of the Belgian team at the Jungfraujoch (Switzerland).

She initiated and led, with Professor F.K. Kneubühl of the Solid State Physics Laboratory of the Swiss Federal Institute of Technology in Zürich (ETHZ), long-term research for the measurement of the energy radiated by the Sun in the far infrared. The sensitive instrumentation developed by Professor Kneubühl's group was flown on board the versatile automatic balloon gondola developed by the Space Research Group of the Observatory of Geneva, led by Professor D. Huguenin. The absolute fluxes thus measured in the wavelength interval from 0.45 to 3.30 microns made it possible to constrain the minimum temperature of the solar photosphere to $4380 < T < 4630K$.

For almost 30 years, Edith Müller has been very active in the International Astronomical Union (IAU) in which she served as Secretary, member of the Organizing Committee, Vice-President and President of various commissions, in particular the Commission on the Teaching of Astronomy and the Commission on the Exchange of Astronomers. In 1973, she was elected Assistant General Secretary,

7

Edith's parents (Courtesy R. Frey, from 1 to 8)

Edith, a few months old

Rosmarie FREY (Wetzikon, Switzerland)
THE LIFE OF EDITH A. MÜLLER :
PERSONAL AND PROFESSIONAL ASPECTS

Edith Alice Müller was born on February 5, 1918 in Madrid, the second daughter of Anna and Max Müller–Niggli. Her parents had already lived in Madrid for four years. Max Müller, son of a post office official from Zürich, had originally been employed by Brown Boveri as an engineer. He founded, in 1919, a subsidiary of that company in Madrid

which was prosperous until the Spanish civil war. The family could afford to live comfortably and enjoyed a good reputation in Madrid. Spain soon became their second home country although the mother had embarked somewhat reluctantly on this adventure. Spain was still a little known country at that time and meant almost the end of the world. Together with her sister Jenny, who was a year older, Edith had a happy, protected childhood. The family moved in 1926 into a house with garden in Pinar avenue, built by the parents in Swiss style. Several employees took care of their physical well–being and belonged to the family. The parents had many friends and their hospitality was proverbial. Apart from celebrations and extensive banquets, their house was always open to travelling foreign guests and homesick Swiss. This way they won many new and interesting friends. For instance, the opera singer Emmy Krüger impressed the two girls when living with them for some time – in particular by her need for tranquillity and then because she would swallow each day a raw egg for her voice. On Sundays the family played games, also with guests – a wink with the eye or a kick under the table were part of the bridge game – and the playing cards would never be missing on picnic outings either. Throughout her life Edith Müller never lost her sense of openness and hospitality.

Her enthusiasm for travelling also goes back to her childhood. The annual vacation in Switzerland offered the opportunity to stay in contact with the home country. Part of the vacation was spent with the parents of Anna Müller–Niggli,

11

whose father had been the head of a teacher's college in Zofingen. Apart from Switzerland, other countries were visited, such as Spain, Portugal, France and England. The family was fond of travelling, always with a lot of luggage. Edith used to count 4 necessaires, 4 wardrobe trunks, 2 hat boxes, 1 laundry bag, 1 golf bag, 1 umbrella bag, 4 coats, 3 handbags, 1 briefcase etc. After all, one had to keep up appearances when staying at the Eden au Lac hotel in Zürich or at the same hotel as the King of Spain. On those occasions she had to practise to bend her knee for the perfect curtsey in front of a mirror.

From 1924 until her maturity diploma Edith Müller attended the German school in Madrid. Her initial difficulties with the multiplication tables – her mother made her work on them during the whole summer vacation – were soon overcome and she then showed so much interest in mathematics that her father asked for her dispensation from needlework in order to increase the number of math lessons together with the boys. A close friend describes the situation at school as follows : *As far as I can remember Edith had been an excellent student, particularly in science subjects.* Many a teacher called Edith to the blackboard when he was no longer at ease with a problem ! Even though zealous and hard working, she never boasted about her abilities and was therefore appreciated by both her teachers and classmates. Of these school years some of the friendships lasted until her death. Spain remained her second home country, and whenever she heard Spanish or could speak the language it would make her happy.

After matriculation the two girls moved to Zürich to take up their studies. Edith Müller enrolled in mathematics, physics and chemistry while Jenny started medical studies.

The parents never questioned their choice which was somewhat unusal for girls at that time. Equal rights was an obvious fact for them ; in fact their mother always considered as utterly unjust their having been denied the right to vote.

Swimming (at the right)...

...cycling...

... Skiing (Edith at the left)

The family would soon be reunited, as the parents had to return to Zürich under dramatic circumstances during the Spanish civil war. Thus Edith Müller enjoyed again the warm support of her attentive parents who shared the tension during exams and the excitement of success. The parents' large circle of friends gave their daughters access to Zürich's society, and they soon had many acquaintances of their own. Excursions with friends and parties were part of their student life.

Edith Müller completed her studies at the science faculty of Zürich University and obtained her doctorate degree in 1943 with distinction. Her dissertation on the ornaments of the Alhambra in Granada underlined again her close ties with Spain.

The Müller family (from left to right : Jenny and Rosmary, Anna, Max and Edith Müller), Xmas 1954.

After several short assignments as an assistant at the department of mathematics at Zürich University, as a mathematics teacher at a girls' high school and as a scientific assistant at the Federal Observatory Zürich, she had the opportunity to work as a visiting astronomer at the University of Cambridge and as a scientific assistant at the Observatory of Ann Arbor. Then followed a one–year assignment as a scientific collaborator at the Institute of astronomy of the University of Basel, after which she moved again to Ann Arbor for seven years. There she was a scientific collaborator at the Observatory of the University of Michigan, with an additional teaching assignment for solar physics during the last two years. She lived happily in the USA and had many pleasant memories. During that time she established many international friendships.

After her return to Switzerland she lived in Geneva from 1962 until retirement. There, after being senior lecturer for astrophysics at the University of Neuchâtel, she was at first lecturer, then senior lecturer and from 1972 ordinary professor of astrophysics at the University of Geneva. This period of time was interrupted by numerous visiting professorships, for instance in Kiel, Lagonissi, Mexico City, Istanbul, Florence, Arizona, Utrecht etc. Her activities introduced her to many national and international organizations where she participated actively and presided over important commissions. Examples are the I.A.U. (International Astronomical Union), ESRO (European Space Research Organization), ICSU (International Council of Scientific Unions), EPS (European Physics Society) and many others. Her scientific work, reflected by a long list of publications, as well as her diplomatic abilities, were appreciated everywhere.

It is therefore not surprising that many relationships survived until after her retirement and she would be invited again and again to give presentations. Her hospitality and her extensive world–wide correspondence showed that she was spontaneous in building relationships which would last. Thus after her retirement in 1983 and having moved to Basel, she succeeded in establishing new friendships : at the association of academic women, the parish, the neighbourhood, when travelling etc. Happy days were often spent with friends in Thusis, Überlingen or while travelling with friends, since travelling remained her favourite hobby. Her apartment was always open to visitors. Political, social, religious and cultural subjects were her preferred topics.

With niece Rosmarie, Xmas 1951

Her family – that is her sister's family since the death of her parents – was important for her and she enjoyed spoiling them. She had moved to Basel to be closer to her sister and brother–in–law. During the last years she looked regularly after her sister who suffered from Altzheimer's disease and tried to analyze and understand the implications of this illness. She took much interest in the life of her niece, husband and children and stayed in close contact with her remaining godchildren as well.

The growing symptoms of old age worried her and she often expressed the hope to die from heart failure. In this light her sudden death, after a happy vacation in her beloved second home–country Spain, appears to be a divine ordinance.

Edith in 1951...

16

—————— Fritz EGGER (Peseux, Suisse) ——————
MEETING WITH EDITH MÜLLER

It was on the occasion of the presentation of her thesis Moresque Ornaments and Group Theory (dealing with the geometric figures of the Alhambra, in Granada) that I met Edith Müller , in 1943. I was to begin my own studies as a physicist, and she impressed me deeply indeed, on one hand by the selection of her thesis subject, mostly of an artistic nature (see hereabove p.2), but treated with such a rigorous weapon as mathematics, and on the other hand, and perhaps still more, by the fact she was facing with determination a purely male and stern assembly who listened politely.

Later on, I had several occasions to meet her again, and to recognize that self-confidence, that enthusiasm with which Edith was facing all her endeavours.

Another characteristic aspect of her behaviour was her versatility ; either she had to introduce to a group of amateur astronomers the most recent results of astronomical research, or she had to organize an extra-mural expedition of the team of her co-workers at the Zürich Observatory, or again simply she wanted to discuss some topical problem with colleagues... ; one could always rely fully upon her kindness and availability.

In the early sixties, she was temporarily in charge of the teaching of astrophysics at the University of Neuchatel. It was under rather difficult conditions, as the interest there was oriented primarily towards astrometry, time and frequency determination. She was bringing us some fresh wind...

Her taste for human relationships led her quite naturally to meet scientists of the international community, in which she took the heaviest duties, at the highest level. Having myself moved rather far from astronomy strictly speaking towards responsibilities in education and in the formation of teachers, I was happy to meet Edith  again at the IAU Commission "Teaching of Astronomy", where she displayed extraordinary effort in order to insure and to prepare the indispensable shift in generations necessary for the progress of our science. At the General Assembly of

kindness and the feeling of pleasure in life. With all kinds of initiatives, she created in the astronomical institute a true social community with a lot of activities going beyond the usual scientific work.

JOSO

The Joint Organisation for Solar Observations was established by Kiepenheuer, Righini and myself during the 1967 Budapest I.A.U. symposium on solar activity. We were of the opinion that efforts should be started and -if already ongoing- united, to search for an excellent site for a common European solar Observatory. Kiepenheuer, at whose initiative the enterprise had been started, became the first president and I was appointed secretary. After Kiepenheuer's death in 1974, I became his successor. The question of the site took almost a decade, and it ended by choosing the La Palma caldera rim called Roche de los Muchachos in the Canary Islands. The place is nowadays known as the La Palma Observatory.

Edith and C. de Jager at the Lunderen Conference on UV spectra of Stars, 1969. (courtesy C. de Jager)

I stayed President of JOSO till 1979, but at that time, I had been elected President of the International Council of Scientific Unions and it became very obvious that there would be too little time for caring well enough for the JOSO, exactly in a period when the site was already defined, and when the next step of choosing the right kind of instrumentation had to be discussed. Several proposals for large common instrumental projects had been tabled; we had to make a choice and secure their funding. Edith was proposed as my successor. She held that job in an excellent way, guiding JOSO through a truly difficult period of its existence.

Ballroom dancing.

Let me end with one very personal reminiscence of an evening during the IAU General Assembly in Brighton, 1970. The participants had the unique privilege of being entertained that evening by the world famous British ballroom band of Victor Silvester. It so happened that I was sitting at a table with the IAU President Otto Heckmann and his wife, both devoted and enthusiastic ballroom dancers, my wife Doetie and Edith Müller. When the music started, the Heckmanns immediately went ahead; Doetie, who has always been a very good dancer - contrary to me- was asked by someone and I remained sitting at the table with Edith. I felt that I had the obligation to ask her for a dance, as convention and courtesy require. At the same time, I realized very clearly that dancing with me would not be a real pleasure for Edith, so : what to do ?

 After a brief period of hesitation, I decided to invite Edith, adding that I was an inexperienced, rather a very bad dancer. That became obvious during the subsequent few minutes. When, finally, after a time span that seemed hours to me, Victor Silvester decided to terminate his melody, I was only too glad to bring Edith back to the table, apologizing and adding : "as you see, I *am* a bad dancer". According to my memory at that time, she confirmed that, and said "yes, you are right".

When, at a much later occasion, I told this story to Edith, she objected far more violently than I had expected or had ever seen her doing before. She insisted that she could *never* have said such a thing. Indeed, what I intended to tell as a brief and actually unimportant funny story was nearly taken by her as an offence, to such a degree that she was close to weeping. For me it was an initially unexpected reaction to something just meant to amuse people, but for Edith the

27

matter was different. Saying bad things about or to people was so much against her very nature that she felt nearly insulted by my little story. I understood that she did not want to be identified with such a rude attitude. In retrospect, I think that my sadness about a failure in ballroom dancing with someone whom I liked so much had made me misinterpret her reaction to my apologies.

My revenge ? Doetie, my wife, had observed my frustration, and a few months later, she persuaded me to take lessons in ballroom dancing. What I had never expected : I became fond of it and we continued our weekly lessons for no less than 24 years during which period I passed all possible examinations, getting silver, gold and diamond awards, and thus became a very devoted dancer of ballroom and latin-american. Most happily, at another dancing occasion, I finally had the good fortune of meeting Edith, and it was at that opportunity that I invited her again to the ballroom, and there we went.

Edith. A wonderful girl.

─────────── Jean-Claude PECKER (Paris, France) ───────────

FROM ZÜRICH TO THE IAU.... LAUGHS AND DUTIES

Chocolates, chocolates, chocolates....The first memory I have from Edith (1948) is one of a smiling generosity, that of a sort of smiling angel coming from a rich country, Switzerland, which had not suffered from the hunger of western Europe.. Remember that in the winter of 1945, the Dutch were eating the tulip bulbs...I was at Utrecht, after the meeting of the IAU in Zürich, where Edith had played an essential part in the organisation. I could not have attended it ; I was myself a very young chick, still working for my Ph.D. with Marcel Minnaert. After Zürich, Edith needed some fresh air. She came through Utrecht ; what for ? I could not remember...For how long ? I just do not know. Her path had the smell of chocolate, the colour of her unforgettable smile, and the sound of her laughs....

Of course, we became good friends, and met on many occasions, ever since... I soon discovered that she was not exceedingly happy in her very subaltern role in the solar physics at Zürich and Arosa. Max Waldmeier's possessive personality did not fit the smiling and imaginative way of Edith!... The situation was worse than only conflicting... She had to leave Zürich and Arosa, to go to a better intellectual climate..

She found it first in Cambridge (UK), and soon after, around Leo Goldberg, in Ann Arbor. I visited Edith in Ann Arbor. That was, I think, the next step. We went once to visit (in her car) the radio-astronomical station of the McMath Observatory (if I remember well its administrative connection), where John Haddock (whom I had met earlier, -1952- in Khartoum, at the time of the solar eclipse) was the key-person. But in her car, we discussed much more a subject of common interest ; namely the abundances of elements in the Sun. We discussed so much, so intensely, on that speedway, that we (I share that responsibility !) missed the side-path going away from the highway to the radio-astronomical side....Nevertheless.... We spent later a wonderful time with the Goldbergs, on the banks of the beautiful lake on the banks of which they had their summer house ; I believe we went even swimming, at least rowing. This must have been in 1954, when I spent some time in McMath, studying, with Helen Dodson, the three-dimensional motions in solar prominences. Edith's little apartment in Ann Arbor was delightful, like all the other places she has been filling with her joie de vivre, pleasant, cosy, warm....

We met at several solar meetings, at several IAU symposia, or IAU General Assemblies.... She was the good fairy of the IAU ! I still remember her first visit to me in Paris, when I lived in Sceaux. She loved to take pictures ; and she took pictures then of my three children ; it must have been around 1957. Since that time, none of her frequent passages through Paris was without some lunch or some dinner, and often a "promenade" through the streets of old Paris, which she cherished. Then we tried to cheer each other up, whenever our life was going through difficult moments. She often called me "pobrecito !" remembering her Spanish education ; and, in return, I called her Alicia....

IAU... At Hamburg, 1964, we attended the IAU meeting , - of course! ... One day of vacation was offered to us, and we visited Bremen. And, together with the Herzbergs, we were wandering around the old town, walking above the walls of the old city, looking at the canals, at the river, at the old roofs, and at the same time speaking of astronomy, of the Sun, of molecules... It was indeed charming! An interesting life was associated with the IAU, then. It was rare that the traditional closing dinner was not followed by some dance. Not in Hamburg however... But in Prague, in Berkeley, in Baltimore, elsewhere, ... so many times , I had the pleasure to dance, one waltz or one tango, with Edith. She was a splendid dancer..Fading remembrances, in the smoke of long ago....

When the time came to choose a new Assistant General Secretary of the IAU, after Contopoulos, in 1973. I was of course, informally consulted, in particular by Kees de Jager. And she was enthusiastically elected. She always considered me (in that capacity!) as her "great-grandfather" (Kees de Jager was my "grand-son", and "her grand-father", in this fantasmatic ge-  nealogy of General Secretaries of the IAU - a genealogy which owed everything to her pleasant imagination!... We worked together on so many problems!..) Each time she came to Paris (even before she decided to put the seat of the IAU in Paris), we had a chance to meet, to chat.

Edith has been the first woman to be the General Secretary of the IAU. She accomplished, in her term, extraordinary performances.

On the one hand, the foreign affairs of the IAU went through a very difficult period. We had hoped for years to meet, in 1982 , in Varna or perhaps in Sofia, as our Bulgarian Colleagues were eager to have us, and as Edith of course was encouraging contacts outside the"majority" flow of the IAU...But at the last moment, i.e. during the General Assembly of Montréal, she learnt, to her dismay, to the dismay of all in whom she then confided, that the Bulgarian hosts felt unable to organize that General Assembly. For days, in Montréal, without being able to attend the Commission meetings or the Joint Discussions, Edith hanging on the telephone wire. She left Montréal having obtained the promise of an invitation in Spain, for 1983 ; unfortunately, it could not be finalized. Finally, Patrick Wayman (who succeeded Edith), with the help the USA National Committee, obtained (only in February 1980) from the kindness and willingness of our Greek colleagues to organize the General Assembly in Patras in 1982. They paid indeed a very great tribute to their kindness!..... Edith, Patrick, and our Greek friends saved us from a difficult situation, but it is Edith who suffered the more....

Another problem was more internal, so to say, to the IAU, but was of lasting consequences. The Secretariat was in Lausanne (Edith being in Geneva), with the same people who were formerly in Nice, in Prague, in Utrecht. ...Edith discovered that it was not an ideal way of functioning; and that the task of the Executive Secretariat was, in a way, affected by such changes from one place to another every three years. Therefore she decided a drastic change of policy, of course with the full approval of the Executive Committee: one knows the essential role of "hortator", as on the Roman "galères". Edith was the smiling and firm "hortator" of the E.C.... She thus moved the permanent secretariat to Paris, in a little house essentially unoccupied, at the gates of the Paris Observatory. It was a nice place; I am not sure that moving from it later to the IAP was a good move: Edith in essence was right to want it independent of the French institutions, and at the same time close to the local astronomers. I wonder whether she has written something on this interesting decision; it would have been no doubt fascinating, in view of the fact it was a disputed decision!...

Thereafter, these accomplishments of Edith, the devoted Edith, the always smiling Edith, have been decisive, in many ways, in the future life of the IAU...I am proud of my great-grand-daughter!

Another circumstance which I shall not forget was the meeting in Ebensee, on the occasion of an IAU Symposium, by this magnificent lake in the Black Forest, debating on the Sun, of course, her lasting love. She was, as ever, eager to bring in the young, and especially when they were in difficult circumstances. I remember an almond trout dinner (those trouts!... Edith liked the good life...), with Judith Pap, then on the verge of leaving Hungary, perhaps for ever, and my wife Annie. She gave some good advice to Judith... Edith was always oriented towards the fate of the young, of those in difficult situation, of the young women in particular. She was deeply involved in the Commission for Exchange of Astronomers, and that for Teaching of Astronomy. After I left the chairmanship of the Chrétien Committee, she took over gently, and very efficiently... So, around the trout of the lake, we spoke about Judith's future, and I know Edith had been instrumental in taking her away from her rather dim future in Debreczen towards a brighter career in the USA... She was proved later not to have been wrong.

Edith, of course, was a welcome visitor to our house on the island, - not an IAU visitor, but just a friendly one. When she left the island, after of few days of enjoyment of the beautiful landscape, rocks and sands, waves and dunes,... we left the island together, for where ? ; and I remember a breakfast in Niort near the old "marché". Pleasant. Then, her train, my trip. We had to separate.

 So many memories, Berkeley-1961, Bilderberg-1968, from IAU to Sun, from Sun to IAU... Many more... Recently still The Hague- in 1994... I did not realize we were to lose her so quickly after that. The smile, the laugh, the enthusiasm of Edith, I will always remember as one of the best aspects of my life as an astronomer. Thanks, Edith, for your laughs, thanks for what you did. Thanks for your everlasting willingness to perform your duties, and to enjoy life !....

──────── Hermann and Mary BRÜCK (Midlothian, Scotland) ────────
EDITH IN CAMBRIDGE AND IRELAND

Edith Müller was a good friend of many, including ourselves.

Hermann knew her from a very early stage, perhaps from the Zürich IAU. In 1951 she spent some time at Cambridge Observatory and also visited Dublin and Dunsink Observatory. Hermann and I were not yet married. I myself happened to be on my first visit to Cambridge while she was there. Dr Arthur Beer kindly took charge of us and showed us round the Colleges. I have the most pleasant memories of those few days.

She then came to Ireland for a week or so. Hermann entertained her at Dunsink and showed her round the city, and I took her on a tour of the country. It was also a very enjoyable time.

We were married later that year (1951) and afterwards met Edith on a number of occasions and always regarded her as a special friend. We particularly remember the Dublin IAU 1955.

Our last meeting with Edith was in April 1994 when she was in Edinburgh for the Royal Astronomical Society/European Astronomical Society meeting. Hermann, who has arthritis, was unable to attend, but Edith very kindly came out to our house to visit him after the meeting, on Saturday, April 9. We look back with pleasure on that afternoon with us, when we talked about old times and about old friends.

With Mary Conway (who was not yet Mary Brück) at Cambridge, 1951.

Bernard PAGEL (Copenhagen, Denmark)

FROM A SEMESTER IN CAMBRIDGE

I started as a research student in October 50 and so was overlapping with Edith Müller's period in Cambridge. What I mainly remember is tea-time conversations presided over by Redman, where the free-wheeling atmosphere of Cambridge was very favourably compared to the working conditions in Zürich under Waldmeier!

Our real friendship started later, at Michigan, and continued up to the end...

THE SUNNY YEARS OF ANN ARBOR

Eager to leave Zürich, Edith, in 1951-52, spent a half-year in Cambridge (UK), then moved to AnnArbor in 1952. She stayed there until 1962, but for a year spent at the Basel University. These were probably the most fruitful years of her life, as a scientist. And she enjoyed tremendously the quiet and peaceful atmosphere of Michigan, with its forest, and its lakes. She made many friends there, - some are no more with us, unfortunately. The Kaplans, to whom common friends recommended her from Switzerland, remember of her arrival there,... and how the life in Ann Arbor was a little bit a discovery of the world.... Donat Wentzel, (who quotes some other witnesses), Günther Elste, Anne Cowley do remember the cosy atmosphere Edith was creating around her....Margaret Burbidge witnessed not only parties on the lakes but the birth of a memorable paper on chemical abundances. Lawrence Aller (the A of that famous GMA paper; – the G was Leo Goldberg, whom we all miss) tells us about the genesis of this paper, one of the most often quoted paper of this period. But Ann Arbor was a beginning : there already, as told to us by Leo Houziaux, she began to get involved with the teaching of astronomy.... Sydney Van den Bergh, who knew her at that time, adds a moving touch to these remembrances, and evokes the right balance that Edith was able to keep between her private life and her scientific interests. On some of the pictures included in this chapter, we see other old friends of Edith, the group in the MacMath Hulbert Observatory for example, with Keith Pierce, Orren Mohler, Helen Dodson, John Waddell.... This was a sunny period indeed !

———————— Wilfred and Ida R. KAPLAN (Michigan, USA) ————————

FRIENDS IN ANN ARBOR

In the fall of 1951, as we recall, Edith wrote to us from Switzerland for advice about coming to Ann Arbor, to take a position in the Astronomy Department of the University of Michigan. She had obtained

Following one such visit, we drove with Edith and friend Mildred Denecke to Stratford, Ontario, to see several plays at the famous theatre.

We supplement these recollections by the following paragraphs, loosely based on accounts of Mildred Denecke and another friend, Elizabeth Boles, both part of the circle in Ann Arbor.

Edith was a warm, open and generous person, who enjoyed people and maintained her friendships. She was also frank in saying no to something that she did not want to do or to participate in, but would give her usually quite logical reasons. She almost missed her Ann Arbor farewell surprise party because her hayfever was bothering her and she could see no reason why she should attend a picnic. The English gentleman who was sent to call for her to attend a "small" get-together had to reveal the surprise and tell her that over 50 people were waiting for her -the guest of honor- before she consented to come.

Quite early in her Ann Arbor years, she bought a house ; she loved it and loved to entertain in it. She started out bravely with very few pieces of furniture ; her first coffee table was a footlocker with a cloth over it. She gradually acquired a full house of furniture, including an unusual sofa. When she returned to Switzerland, she took both the long sofa and her American refrigerator (at a time when Swiss ones were very small). The house was the site of many enjoyable holiday parties, suppers and picnics. Her Christmas-New Year's parties were always special. Her Christmas tree was memorable because it was lit with real candles (rare in the US) ; however, they were always extinguished before the guests adjourned to the basement for dancing.

Because of her scientific background, Edith was always well organized, even in her private life. She knew how much she could do and planned accordingly. On one of her visits to the US to give a paper, she first worked on her notes for the paper and then said : "Now I'm ready to have fun" and then joined friends for sight-seeing. She usually programmed a nap or a quiet cup of tea sometime each afternoon. Friends were always amazed that when she was having dinner guests, she would have everything ready and then lie down for a rest for about a half hour before the guests were expected. She also kept a card file recording who they were and what she had served ; she never served the same food twice to any guest.

Edith liked nice things. Her clothes were always of the best materials. Mildred writes : "During one of her visits to the US to observe at Kitt Peak, I joined her for a long weekend. After crawling around the McMath Solar Telescope for a while, we went off shopping. She had been invited to speak at the opening of the Cerro Tololo telescope in Chile and was planning her wardrobe for the event. We explored several of the fine jewelry shops in Tucson before she found just the right silver and turquoise Indian necklace to go with her simple but elegant dress". Edith was known to use her time while speaking on the telephone to do some sewing ; in this way, she made a beautiful wool suit.

About 1976, she went with a group of eight friends (including Mildred and Elizabeth) to Mt. Sinai, to accompany the photographer Fred Anderegg, who made fine photographs of the icons and mosaics at the ancient monastery there. The last stage of the journey was made in an open truck over a rough road. When the gates to the monastery were closed, one could enter only by being carried up in a large basket. During their stay, the visitors slept on straw mattresses on the floor.

Edith took good photos with her little camera. When she had a grand view to record, as from the top of Mt. Sinai, she took a series of overlapping pictures to form a panorama. She became very good at this. And she was always the first one to have her pictures mounted in an album, clearly marked and dated, as well as the first to send appropriate copies to her friends.

When she returned to Switzerland, her life gradually changed. She no longer had her little house in which to entertain ; besides, such home entertaining was not as popular there as in the US and was less called for in view of the marvelous restaurants and patisseries in cities such as Basel and Geneva, where she lived. Nevertheless, visits to Edith at her apartment were always delightful. For those who stayed overnight, there were always small bars of Swiss chocolate on the night table in the guest's room, and she was always able to put together an excellent light supper from "nothing". In Basel, she proudly showed friends the zoo ; on one occasion watching with them a new baby elephant for a long time before going in for tea and cakes.

Edith was not "into" computers, but in her later years had discovered TV. She would never become a slave to the tube, but Mildred recalls a pleasant evening hour with her spent watching a mystery drama.

—————— Donat G. WENTZEL (Rockville, Maryland, USA) ——————
WITHIN THE ANN ARBOR COMMUNITY

I shared an office with Edith and, as a newly baked physicist, I learned from her much of the workings of the astronomical community. Our office was built into the old observatory's entrance foyer and was very publicly accessible. She was her usual friendly and bubbly self with everybody who visited, both students and faculty. She was always interested how everyone was doing, especially how the students progressed in their studies.

Ann Arbor : a relaxed garden party. From left to right : C. Cowley, Edith, X, Peter Boyce, H. Dickel. On the front, right : D. Wentzel. (courtesy A. Cowley)

In Ann Arbor, as everywhere...; the tea was always ready for the friends... (courtesy R. Frey)

Within the department, Edith was one of the most helpful and supportive members of the staff to the graduate students. She was always willing to discuss or advise, and her upbeat and cheerful manner made us all feel better. For the women students, she was a special inspiration, showing us that unequal treatment in jobs doesn't mean you can't be a world class scientist. In Ann Arbor, Edith had a multitude of friends, both astronomical and in the community. She loved her home and garden there, but we were not surprised when she accepted a highly deserved professorship in Switzerland. It was unfortunate for the University of Michigan that they did not have the sense to offer her a faculty position. Her outstanding research and service to the IAU which continued over her entire lifetime are a great credit to her memory.

――――――― Margaret BURBIDGE (La Jolla, California, USA) ―――――――
FROM A SUMMER SCHOOL IN MICHIGAN.

Geoffrey and I first met Edith Müller during the famous Summer School in Ann Arbor, Michigan, at which George Gamow, Walter Baade, Edwin Salpeter, and George Batchelor were lecturing. During this Summer School the participants were taken to the McMath-Hulbert Observatory where Edith was working, for a day of fun, a picnic by the lake, and the opportunity to swim in the lake. Geoffrey and Allan Sandage confused the Observatory staff by exchanging their name labels, but there was retaliation during the games on the lake, when some of us, astride floating logs, were having mock battles in which the "ships" attempted to overturn one another and dump the occupants under water. Geoff was at that time wearing his hair quite long, and Edith and Ed Salpeter attacked him with a large pair of scissors and tried to cut his hair. Geoff claims he did not get very wet ; my recollection is that those of us already in swim suits came out of it the best.

During her years at the University of Michigan, Edith had been working on the monumental study of the abundances of the elements in the sun, which culminated in the 137-page paper by L. Goldberg, E.A. Müller, and L.H. Aller (1960, ApJS,**5**,1). Abundances of 42 elements relative to hydrogen were determined and published in this paper. The data were available a few years before the 1960 publication, and were an indispensable source of information during our work on the synthesis of the elements in stars.

In the early 1970's I had many interactions with Edith at the IAU, especially during my term as President of Commission 28 (Galaxies). Through her work as General Secretary of the IAU and on its Executive Committee, and perhaps especially as President of Commission 38 on the Exchange of Astronomers, her contacts were world-wide. She was loved by all her many friends, and we looked forward with pleasure to the chance to talk with her, over a meal, over a cup of coffee, or for a peaceful moment during the rush and tumble of international meetings. I treasure especially the memory of one particular long walk we enjoyed along the shore of the lake in Geneva, the dinner we ate together, and the talk we shared on astronomy, astronomers, the world in general.

It is an honour to have known and admired Edith and to have been among her many friends. We miss her.

Friends in Ann Arbor : Edith with S. Segré at her left, and other friends. (courtesy M.S. Vardya)

——————— Lawrence H. ALLER (Los Angeles, USA) ———————
AROUND THE CHEMICAL ELEMENTS.

From the 1950's to 1962, the Observatory of the University of Michigan in Ann Arbor was graced by the presence of Edith Müller who worked with Leo Goldberg and me in a study of the chemical composition of the sun. (*) Before this investigation was undertaken, various attempts had been made to modernize determinations of the abundances of the most abundant elements in the solar atmosphere, but there had been no attempt to update Henry Norris Russell's classical work of the late twenties taking advantage of the progress that had been made both in the study of stellar atmospheres and atomic data. A few primitive attempts had been made to analyze stellar spectra (e.g. of B stars) trying to take into account the variation of temperature and pressure with depth. The standard mode of procedure for the sun and stars was to use the curve of growth technique for an isothermal atmosphere at a constant pressure. At Michigan, we introduced the refinement of trying to take the effect of the solar atmospheric structure into account via the method of weighting functions (Unsöld, Minnaert, Pecker). The model atmosphere adopted was based on the work of Aller and Pierce (1953) and extended by Elste to deeper and higher layers. Equivalent widths were taken from investigations by C.W. Allen, the Utrecht work and measurements made at McMath-Hulbert Observatory.

The greatest difficulties arose from the sorry state of the then-available oscillator strengths or f-values. Since the equivalent width (often called the "intensity" of the line) depends on the product Nf where N is the number of atoms capable of absorbing the spectral line in question, the abundance will depend on the accuracy of the f-value no matter how accurate our analysis. These f-values may be calculated by theory for simple atomic transitions or measured by experimental procedures. Iron turned out to be the "bête noire" of the abundance problem. The equivalent widths of the strong lines, for which f-values might be easiest to measure, are affected by line broadening processes so that the observed values involve not only f-values but collisional broadening processes so that the observed values involve not only f-values but

---

(*) The Abundances of elements in the Solar Atmosphere. Leo Goldberg, Edith A. Müller, and Lawrence H. Aller, Astrophys. J. Suppl., n° 43, 1960, **5**, 1-138.

collisional broadening parameters. Lines of Iron, one the most abundant metals, dominate the solar spectrum. Many metals present less formidable problems and their abundances were established reliably first.

A great impetus to the accurate quantitative analysis of the solar atmosphere came from studies of the carbonaceous chondrites and hypotheses of the origin of the solar system. A critical point was whether the non-volatile elements in these meteorites had the same abundance distribution as in the sun. The Fe/Si ratio, for example offered one crucial test.

Ann Arbor. Edith and Marcel Minnaert, speaking certainly about chemical abundances !... (courtesy M.S. Vardya)

In connection with the production of his compendium on the solar system (a parallel series to his Stars and Stellar Systems) G.P. Kuiper asked Leo Goldberg and me to prepare an article on the chemical composition of the solar atmosphere. It became apparent to us that a patch-up would not do ; a comprehensive review of the problem was required. Goldberg had heavy administrative responsibilities, not only as head of the Michigan astronomy department but on the national scene as well, wherein he was deeply involved in plans for a National Radio observatory and also in the initial planning for Kitt Peak. As for me, I had the primary responsibility for the bulk of the graduate teaching program and the guidance of graduate students. Fortunately, Leo had met Edith Müller at the 1948 Zürich IAU meeting and was able to secure funding for her to participate in this endeavor.

Thus Edith came to Ann Arbor and took over the bulk of the detailed analysis and assessment of the vast quantity of observational data. She pointed out where new observational measurements were needed and fought through the confusing mass of discordant and sometimes downright contradictory f-value determinations at hand. Through all of this ordeal, Edith maintained a cheerful optimism and exemplary perseverance in the face of daunting frustration. The task could not have been completed, at least in a reasonable time scale, without the capable participation of Edith Müller, who worked so diligently and faithfully.

Although the effort provided a background and inspiration for much subsequent solar abundance work, the study was hampered by poor f-values. As this problem has been cleared up over the years, we now know that the solar abundance pattern (for non-volatile elements) conforms closely to that of the carbonaceous chondrites.

It was a great privilege to have known Edith Müller and to have been associated with her in the University of Michigan Astronomy Department for a number of years. Her friendship was treasured by all who knew her ; her linguistic skills and social graces will long be remembered.

 Léo HOUZIAUX (Mons, Belgium) ────────

FROM ANN ARBOR ONWARDS...

Ann Arbor, July 1958. The times were rather uncertain. The Xth IAU General Assembly was to be held shortly after in Moscow. Because of disagreements between East and West powers over Lebanon, US Marines landed in Beirut and it was not sure at all that the US astronomers would be allowed by the State Department to fly to USSR. As I was at that time touring Mexican, American and Canadian Observatories after one year of study at Berkeley, I decided to wait for the situation to be settled before returning to Europe ; as the head of the American delegation was Leo Goldberg, I planned to end my tour at the place where he was working. This is why I found myself seated at an evening party with Aller close to a young lady who was speaking French with her left-hand neighbour, while she was chatting in Spanish with her right-hand partner, talking either German or English with other people in the room. This was quite a change for me for, during one whole year, I had hardly met in California anyone who had been thinking that one could speak another language than that of their fellow citizens. At the end of the night-gathering, I enquired about the name of that person with whom I had spent a couple of pleasant hours : I had met Edith Müller, and I did not of course realize that I would have so many opportunities to see her again under diverse circumstances.

We saw each other again at a number of meetings, mainly at IAU General Assemblies in Berkeley, Hamburg, and Prague. However, in 1964, Edith became vice-president of Commission 46 for three years, and then remained President for six years. Our common interest in the teaching of astronomy led us to write to each other more often. As a President of Commission 46, Edith undertook initiatives that will certainly be recalled in other contributions of this memorial ; she prepared thorough reports of the development of astronomy teaching. She was very pragmatic in solving all kinds of problems. In 1973, no national contribution having been received from the US delegate for her presidential report, she actually wrote the report for the US herself, her name appearing under "Switzerland" and "United States". At that time, she succeeded in getting funds from Geneva University to start a new IAU publication called Astronomy Educational Material. At the very beginning, this was really a huge entreprise, since one had to enquire all over the world about the existing books, slides, tapes for astronomy teaching.

I still remember her sitting in her new office at Sauverny with an endless list of popularising brochures written in Chinese and other oriental languages, not knowing really what to do with this cumbersome material-naturally of little use for teachers all over the world. This project  revealed itself to be quite useful and remained for about 20 years one of the main achievements of Commission 46. I regretted for a long time not having been able to travel to the Sydney IAU when she organised a wide discussion on the goals of Commission 46 with so many interesting contributions. Having been appointed as a Visiting Professor in Geneva in 1969, I had several occasions to discuss with her problems related to stellar atmospheres ; she worked actively with her graduate students in the field of the physics of solar-like stars, continuing with the most efficient physics of that time the studies she had developed in Ann Arbor and which culminated with the famous "GMA" paper published in the Ap. J. Supplements and which remained for a long time the classical reference in the field of solar abundances. As I served later on as a jury member for Ph D's and other refereeing bodies at Geneva, we had the opportunity to meet at her place and enjoy together during the mild summer in Suisse romande wonderful evenings at the "Creux de Genthod", a place she liked very much to invite friends for outdoor dinners.

We remained in touch during the years where she acted as General Secretary of the IAU, being myself an IAU representative to various ICSU committees (Committee for Teaching of Science, Costed,...) and handling as a secretary to the Belgian National Committee for Astronomy various matters concerning Belgian membership, and financial or institutional matters which, for this country, were already unbelievably complicated. Once, she was somewhat desperate with such time-consuming and unrewarding matters that she called me at the office, asking me : "Oh! Léo, please do something to help me out ; it took me over six months to finish the negotiations with the Chinese, are we now going to start the same business with the Belgians ?" Everyone connected at that time with IAU during these years knows what she really did to solve a number a difficulties and support anyone needing help by making the most efficient use of the IAU funds. Her experience and diplomacy were deeply appreciated by the IAU authorities and she acted as an excellent adviser for the organisation of Schools for young astronomers or for IAU travel funds in the framework of Commission 38.

After her retirement, we had fewer opportunities to see each other, although we frequently saved one lunch-break or one evening to share one. The last time we had lunch together, I happened to carry my camera with me and took a picture of her. This was during the Baltimore IAU General Assembly.

I happened to meet her in the hall of the Congress Centrum in the Hague during the 1994 IAU General Assembly. I was not going to post an announcement looking for people who might be interested in astronomical journals and publications I had to get rid of because of my next retirement. She immediately replied that she too had a lot of books and especially older IAU publications to give away, and she asked me to send her names of interested individuals or institutions ; this attitude is really typical of her character - trying up to the very end of her life to help colleagues and especially young people. Edith will be remembered by a large crowd of astronomers and friends for her bright contributions to science and as an organiser of unusual ability.

At one of her last IAU General Assemblies, the ever smiling Edith... (Baltimore, 1988) (courtesy Leo Houziaux)

——— Sidney van den BERGH (National Research Council, Canada) ———
THE RIGHT BALANCE...

I mostly remember a very kind and dedicated person who seemed to have found the right balance in life between work and her other interests. I think that my late wife, who (as you may remember) died about ten years ago, got to know her much better than I ever did. I still remember that Gretchen and Edith spent a very pleasant afternoon exploring the "grachten" of Amsterdam together.

BACK TO SWITZERLAND : GENEVA...

1962 : Edith, who was by now a well-known astrophysicist, is invited by Marcel Golay, the Director of the Observatory in Geneva, to join the staff. She now teaches astronomy at Geneva and later, also at Neuchatel, and creates a school... She travels a lot; she teaches here and there, at the invitation of her colleagues in Germany, Spain, Belgium, or the Netherlands... A team of "three musketeers" as they used to called themselves gathered around her; Yves Chmielewski, Ramiro de la Reza, Felix Llorente de Andres form around her a real solid team of friends, and they do remember dearly their astronomical initiation by Edith. The abundances continue to take much of her time : Bill Livingston has witnessed her activity at Tucson, in measuring the Lithium abundance with Jim Brault, a problem attacked also with the "musketeers"... From that period also is dating the remarkable series of balloon experiments , during which, together with her colleagues Prof. Kneubühl and Prof. Huguenin,she measured a remarkable variation during the solar cycle, of the solar energy output when measured in the infrared. But as Edith was travelling much, she then met several other scientists. Jean-Pierre Swings reminds us of the occasional meeting of solar astronomers from Belgium and Switzerland at the solar Observatory of the Jungfraujoch. Jan Stenflo tells her about the birth of JOSO, still alive, as shown by Brigitte Schmieder. Emilio Alfaro tells about the rebirth of astrophysics in Spain, which excited Edith Alicia so much... The Dumont's have visited Switzerland with her, ... but missed her in the Canaries; and old friends from Ann Arbor, William Howard, and his wife, so he tells, visited her in Switzerland..., and received her visit in Virginia, where she admired the beautiful free birds...

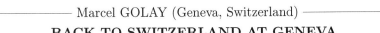

Marcel GOLAY (Geneva, Switzerland)
BACK TO SWITZERLAND AT GENEVA

I was nominated Director of the Geneva Observatory and full Professor of Astrophysics at the University of Geneva in 1956.

Some years later, the Geneva Observatory became the Astronomy and Astrophysics department of the Faculty of Science. Little before 1956, professor Wilhelm

table in front of me, some pages were just slightly raised by the wind, allowing me to catch sight of the announcement of Edith Müller's death, carried off by a heart attack.

—————— Yves CHMIELEWSKI (Sanverny, Switzerland) ——————
THE PRIVILEGE OF HAVING BEEN EAM'S COLLABORATOR

Having known Edith Müller for many years, one of her most striking character traits was her particularly well developed sense of sociability and her cosmopolitan way of life. This many-faceted nature represented a unique blend of the German and Hispanic cultures. That situation was the logical result of the events in her early life, when her parents moved to Spain and from there maintained regular contact with Switzerland.

Edith grew up in a well-to-do, well considered family and thus naturally acquired much savoir-faire, strengthened by her knowledge of many languages. She spoke fluently German, Spanish, English, French and had even some elementary knowledge of Italian and Dutch. She was specially fond of "Schwytzerdütsch", the so peculiar Swiss-German dialect (actually, her mother-tongue). One could immediately tell when she was speaking Schwytzerdütsch on the phone because her bursts of laughter would be stronger and clearer...

The important part of her career was spent in Ann Arbor in the United States, and that obviously had a strong influence on her, also adding to her open-mindedness. It was during this period that she established close ties with those colleagues of her generation who would enter the Gotha of Astronomy.

It was a pleasant and rewarding experience to have been first her student and later her assistant. I shared this privilege with my dear friend Ramiro de la Reza and, later, Felix Llorente de Andres : we formed, together, a nice "gang". She was a good "boss". Although she had an aristocratic sense of professional and social hierarchy, she was easily approachable and we had a friendly, informal relationship with her ; she was always available for scientific discussions and advice. She left me a great deal of freedom and initiative in my work. She had, however, to struggle against my excessive meticulosity and my inclination towards overdiversification to bring me back on the right track, sometimes with success, but in the end a lost battle !

At Sauverny, Edith and her "three musketeers" : from left to right : Felix Llorente de Andres, Yves Chmielewski, Ramiro de la Reza. (courtesy Y. Chmielewski)

She loved to travel, and she did a lot of that, most of the time combining business and pleasure. Back in Geneva, she would burst with a profusion of ideas collected from the many discussions she had during her travels. And, whenever astronomers from abroad would visit the Geneva Observatory, she managed to let us, the young assistants, develop closer and less formal relations with them by organizing parties at her home. Her small appartment was quite cosy and she was as good a hostess as a cook. The atmosphere at these parties was always extremely friendly and lively. They remain quite memorable.

Beyond her work or travels she led a simple and quiet life, giving greater importance to relations with her family and a restricted group of close friends. She drove a funny little car, a 4-wheel "Vespa", which was still smaller than a Fiat 500! I remember once, as a joke, the employees at the mechanics department of the Observatory were able to fit the car into the 2.7 meters-deep elevator and park it on the roof of the building. Edith obviously went through moments of despair when she found out that her car was missing from the parking ground, but all ended with a good laugh...

Her main scientific interest has always been the physico-chemical description of the atmosphere of the average "quiet" Sun. Her part in initiating and interpreting the measurements of the solar brightness temperature in the infrared has been of great significance; these data were of utmost importance for constraining the value of the temperature minimum of the solar photosphere. For the remaining (and main) part of her studies of the atmosphere of the Sun, she was striving to derive as much information as possible from solar spectra of the highest available spatial and spectral resolution and quality. For that purpose she made frequent observing runs at the McMath solar telescope at Kitt Peak National Observatory. In my opinion, what should be stressed about the originality and quality of Edith's scientific work is her emphasis on the systematic use of the physical information contained in high accuracy center-to-limb observations of the solar spectrum, such as made available by the solar telescope of KPNO. The center-to-limb data permitted to check the consistency of the physical description of the line formation and in many cases showed that refinements had to be added to the traditional methods to ensure accurate and reliable results. Examples of such refinements are the use of additional continuous opacity sources in the ultra-violet part of the spectrum or a detailed account of the effects of departures from the assumption of Local Thermodynamic Equilibrium. We may indeed regret that, apart from Edith, so few investigators really made the effort to try and meet the strong constraints imposed by these beautiful center-to-limb data.

Edith used them mainly for the determination of the chemical abundance of elements showing only few lines in the solar spectrum such as Lithium, Beryllium, Oxygen, Potassium, Barium and others. She also attempted to use the information contained in the line profiles (even for faint lines) to better account for the effects of velocity fields in the solar atmosphere. She was among the first who did not assign the non-thermal velocity broadening entirely to a micro-turbulent contribution (which affects the line transfer) but also allowed for a contribution from a macro-turbulent velocity field (not affecting the transfer). From studies of the profiles and asymmetries of selected model test lines at various locations on the solar disk she tried to introduce empirical models of depth- and angle-dependent average velocity fields. Unfortunately, these investigations were not always adequately publicized.

These lines summarize what I have known of Edith A. Müller. Others will hopefully give a more detailed account of her intense activity within the International Astronomical Union. Edith was an astronomer, but she was in particular a Woman-Astronomer, which was uncommon during that period of time. I know from many situations how much she felt concerned with the condition of women in the "hard" sciences, and especially in astronomy. She had lived in the United States during the fifties, where women were mostly excluded from those positions where decisions were taken, and in some cases even from the observing sites themselves. She was a militant in a few organizations for the improvement of women's condition. Actually, I personally have not known much about these aspects of her life, and anyhow it would be more appropriate if a woman astronomer could be found who knew more about that situation. To my knowledge such a testimony has not yet been found, which is quite unfortunate.

Goodbye Edith, it was nice knowing you!

from Kitt Peak. Of course the raw data were conserved by Jim Brault in Tucson. We don't know how but Erik lost all the spectra. For days we searched in all the corners of the Geneva Observatory. I remember very well her bad mood. At the maximum of despair, Erik went to the Geneva city dump! Of course no luck, the Lithium spectra were not there among the onion and patatos peelings. New copies arrived later and we were ready to work. The results appeared in 1975. Some years later, Holweger made a small improvement of the NON LTE results but in fact, the original results were almost the same. These results are valid and maintained up to today, more than twenty years later.

I worked with Edith for ten years. Those were very nice years in which I worked with a lady, a scientist and especially with a fantastic human being. I learned with her the main principles of scientific responsibility. Together with I. Chmielewski and F. Llorente de Andrés I am proud to be one of the three musketeers, one of her three spiritual children.

—————— William LIVINGSTON (Tucson Arizona, USA) ——————

AT TUCSON, LOOKING AT THE SOLAR SYSTEM

Although I never worked scientifically with Edith Müller, she spent considerable time here in Tucson in the early 70s and we became good friends. How could it be otherwise? Someone who always inquired about your work and displayed so much interest and enthusiasm for what you were doing.

This was the time when she and Jim Brault did their seminal analysis of lithium and its isotopes in the solar photosphere. Recently Jim confided that he considered this research period done with Edith and her students to be the best and most satisfying of his career. But Jim may comment on that.

As an American I have always been impressed with the linguistic dexterity of Swiss and I held Edith in special awe. She was fluent in French, Italian, English, Russian, German, and I suppose Swiss-German. I was studying Russian

Professor in Kiel 1964-1965 ; extraordinary Professor in Geneva 1964-1972 ; invited researcher at Kitt Peak National Observatory 1967, 1968, 1969 ; invited Professor in Utrecht 1970-1971 ; full Professor in Geneva 1972-1983 and honorary Professor from 1983. As a teacher, Edith has contributed to forming a large number of scientists throughout the world and maintained high-level scientific cooperation with these colleagues, many of them remaining her friends over the years.

The scientific achievements of Edith Müller are remarkable. She was a leading specialist in astronomical spectroscopy, solar and stellar atmospheres, and with great enthusiasm she regularly observed at Kitt Peak. Among her major results we must mention the accurate determinations of solar abundances in cooperation with L. Goldberg and L. Aller. This work became the basic reference for elemental abundances in the Universe and was among the most quoted papers in astronomical literature for about a quarter of a century. Also, with R. de la Reza and E. Peytremann, she made the first reliable estimates of the lithium abundance in the Sun and their result was of great cosmological significance. Edith Müller was asked to give an impressive number of invited talks and reviews of her research in international conferences. She also contributed to the development of European solar space research. In particular, she participated in many space experiments on the infrared radiation of the Sun and her work led to new results on the outer temperature structure of the Sun. Due to her broad experience she was involved in much international cooperation and gave courses in several foreign Universities, such as Kiel (Germany), Liège (Belgium), Granada (Spain), Sofia (Bulgaria), Istanbul and Izmir (Turkey), Mexico and Utrecht (Netherlands).

In addition to her research and teaching activities, Edith made major contributions in scientific management. Also, besides her work as General Secretary of the IAU and in the Executive Committee of the IAU, she participated in many committees and chaired a large number of them, and it would be difficult to draw a com-

plete list. From 1967 to 1973, she was President of IAU Commission 46 on the Teaching of Astronomy, a member of the Executive Committee of the European Physical Society 1977-1985, President of the Swiss National Committee for the

IAU 1979-1985, President of the Joint Organisation for Solar Observations, President of the Henri Chrétien Fund 1984-1988, member of the International Scientific Committee for the Astrophysical Center in the Canary Islands from 1984, and Associate of the Royal Astronomical Society (UK) from 1979.

 The scientific community, her colleagues and her friends recognized Edith as a great Professor, Astronomer and Organizer. Her human qualities, kindness and availability for her students and colleagues, her pioneering work as a woman in the field of Astronomy, her open mind and courage were unanimously appreciated by her very many friends at the University of Geneva and in the astronomical community in Switzerland and abroad. We will all miss Edith very much, no longer hearing her clear laugh in friendly discussions with colleagues.

——————— Anne B. UNDERHILL (Vancouver, Canada) ———————

ONE GREAT LADY ASTRONOMER.

I first met Edith Müller in connection with my studies of the theories and observations of the spectra of massive stars and her related studies of the spectrum of the Sun. Later I was well aware of her work as General Secretary of the International Astronomical Union. I visited Edith twice briefly when she was living in Switzerland, first in Geneva where she introduced me to the setting of that city and later when she was living in Basel. I recall with interest the small paper-making factory which Edith showed me in the Basel area.

Edith was a kind, friendly person and a very good administrator, witness her work (during 6 years or more) for the International Astronomical Union as General Secretary. I envied Edith her facility with speaking, understanding, and writing several European languages.

On the river, between Hamburg and Ludwigshaven, IAU G.A. Hamburg, (1964). (From left to right : Flora MacBain-Sadler, Nel Splinter (then "Miss IAU"), Edith).

I myself find that speaking and writing in any language except English is difficult. You should remember that I was born in 1920 and brought up in a suburb of the important port city of Vancouver, B.C., Canada. We had small minority populations at that time of Chinese, Japanese, Hindu and Sikh peoples. They spoke their own languages at home, but in school and in business life, they used English in order to communicate with others. Certainly the children from these groups whom I encountered in the Vancouver Public Schools behaved and dressed as closely as possible like we Canadians of English or Scottish descent. In the 1920s and 1930s the only language one ever used was English. Although Canada was, in principle, a bilingual country, French was not used on the Pacific Coast near Vancouver at that time.

For several years I studied French and Latin, the only two languages offered by the Vancouver School system and became quite proficient at each. However, when in 1945 I moved to Montréal to work on a project in physics for the Canadian National Research Council division there, I found that my French bore little resemblance in its spoken form to the dialect spoken in the Province of Québec Later I found out that my French was a reasonably close version of Parisian French. At the time I am speaking of there was no TV and the radio stations in the Vancouver area carried programs only in English. At the present time (1996) radio programs are broadcast in Vancouver in most of the languages spoken by the major immigrant groups who now live in the area.

Because of my lack of experience when I was young of hearing and using European languages, other than English, I regarded Edith's facility with European languages with awe and amazement. Being a good administrator was a valuable skill which Edith possessed in addition to her skill with languages. These skills, in my eyes, made her a person of special value.

―――――――― John T. JEFFERIES (Tucson Arizona, USA) ――――――――
A LEGENDARY POLYGLOT... AND LOVELY DAYS

My first memory of Edith Müller is as the legendary woman who sported at least seven colors on her IAU name badge, one for each language she spoke (and each, seemingly, like a native). This was at the 1961 San Francisco IAU. We had not met formally though I certainly knew who she was through the work on solar abundances of the Michigan group of which she was a leading member. I suppose she knew who I was, too, since she seemed to know literally everyone (and more remarkably to remember their names).

I recall wistfully and with deep fondness the many occasions that we met in later years and the pleasure that I derived from knowing Edith, a pleasure that I shared with her legions of friends in countries all around the world. I believe that this widespread love and admiration was the natural product of her innate kindness, her sincerity and concern for all whom she touched, and for the infectious gaiety which lightened every encounter with Edith.

A solar meeting at Varenne (1960),
with, from left to right : Edith, G. Righini,
Annelie and Eric Forbes. (courtesy G. Elste)

I remember especially well an occasion during a meeting in Nice, after the 1976 Grenoble IAU, when the sea, sky, and mountains combined to make an irresistibly lovely day. Long sequences of barely intelligible talks, presented in lifeless monotones, had taken their toll and a drive in the hills above Nice seemed infinitely more appealing than yet more of that punishment. I had a car (Frank Orrall, Jack Zirker, and I had rented one to drive from Grenoble to Nice and that is another treasured memory) so Edith and I decided to drive to Vence to pay homage to Matisse, to visit his Chapel if we could (which, as it turned out, we couldn't), to walk quietly in the lovely village, eat some lunch, drink a bottle of wine and spend what remained of the afternoon at the Picasso museum above the ocean at the cap d'Antibes. It was indeed a golden day, stolen from the strict demands of discipline, but infinitely richer, both then and in memory, than the day at the meeting could possibly have been.

The last time we met was in 1988, in Paris on the occasion of Jean-Claude Pecker's retirement. Edith was as vital as ever and as deeply engaged with the wide community of her friends and protégés around the globe. It is hard now to accept the fact that she is no longer with us : with all those to whom she bound herself so closely I must be content with the memory of this fine woman, as I am grateful for my good fortune in having known her.

George CONTOPOULOS (Athens, Greece)
FROM ONE GS OF THE IAU TO THE NEXT...

Edith Müller was the Assistant General Secretary of the International Astronomical Union during the period 1973–76, while I was the General Secretary.

Her help during this period was really invaluable. She took care of all the Symposia and Colloquia of the IAU, of its Publications, and many other duties during a period of a growing IAU, with thousands of members, 50 Commissions and about 50 adhering countries.

Besides the yearly Meetings of the Executive Committee of the IAU we had two more meetings of the Officers (the late L. Goldberg, President, Edith Müller and myself, together with our faithful secretaries Arnost Jappel and Jarka Dankova) every year. These meetings took place in Geneva, or in Arizona, or in Greece. Whenever we had our meeting in Geneva, Edith was the host. All the matters of the Union were discussed at these meetings, that filled completely a whole day. I remember Leo Goldberg arriving after a sleepless night on the plane, completely exhausted at the end of the day. But then Edith would take us to dinner at her home. I always remember the wonderful "fondue" that she was preparing for us.

The period 1973–76 was particularly difficult for me, because it started while Greece was under a strict military junta. But I had the continuous support of Leo Goldberg and Edith Müller, who often replaced me whenever I could not travel abroad. Fortunately, at the end of my term, the junta was over, and I was more free to move, especially while preparing the 1976 General Assembly of the IAU in Grenoble. I always had the help of Leo and Edith, and of our excellent French hosts. The Grenoble meeting was a nightmare with 200 meetings and innumerable problems. I said to my family : *"If I survive this meeting I will never be afraid of anything"*. But all went well, and as I was passing over the General Secretariat to Edith at the second session of the General Assembly, I said : *"I am the most happy person of the General Assembly at this moment as my task is over"*, and *"I am quite sure that I leave the work of the IAU in very competent hands"*.

But that was not the end of the day. Edith invited the new Executive

Committee for a meeting the same afternoon, in which I was only a Consultant, but still needed to relate my experience. The meeting lasted several hours and my family in the hotel were alarmed because I had disappeared without leaving a notice where I was going. But this meeting, and the whole period of 1976–79, when Edith was General Secretary, marked a new advance of the IAU.

My collaboration with Edith did not stop there. She was kind enough, later, to supervise closely the thesis of a student of mine, Miss M. Stathopoulou, who worked in solar spectroscopy, a field closely related to Edith's interests. Edith provided not only scientific advice but also warm hospitality to her, and to all foreign students. Thus, I always remember Edith as a friend that would provide her help generously, and a wise collaborator that would give her advice and support when this was necessary.

--------- Patrick A. WAYMAN (Dublin, Ireland) ---------
EDITH MÜLLER, GENERAL SECRETARY, INTERNATIONAL ASTRONOMICAL UNION.

Having joined Cambridge University Observatory as a graduate student in 1948, I first heard of the International Astronomical Union from those who had returned with enthusiasm from the Zürich General Assembly of that year. It became possible at Zürich for former colleagues from before the prolonged period of the Second World War, 1939-45, to meet again in amity and look forward to renewal of astronomical work as an international enterprise and to its future development in whatsoever way might be contrived.

Edith Müller had worked hard on behalf of the Local Organising Committee to ensure the success of that meeting in Zürich and it was through friendships begun then that she was able to carry out research in solar and stellar physics in Cambridge and in the United States and so establish herself as an authority in her chosen field. It is easy to be convinced that Edith's adherence to the value of international cooperation began at that time, continuing through to the very end of her life. Her interest in the international aspects of astronomy reached a peak during the years 1976 to 1979 when she served as General Secretary of the International Astronomical Union.

Eminent personalities at Prague (IAU GA 1967), from left to right : J. Oort, Edith, J.C. Pecker, Lubos Perek, from back, P. Swings.

Although I had for long known of E.A. Müller by name, she first came to my notice in person in Prague in 1967 and again in Brighton in 1970, but I was not present in Sydney in 1973 when she was made the new IAU Assistant General Secretary ; she was to become the first woman to be chosen to follow on as General Secretary of the IAU.

So it was that during 1975 I was approached by Leo Goldberg, then President of the IAU, as to whether I would be prepared to undertake the work of Assistant General Secretary from the time of the 1976 Grenoble General Assembly, with the expectation of taking on the more onerous responsibility of General Secretary from 1979. This would be in succession to Edith Müller and would be the time when the IAU Secretariat would be established in Paris as a permanent locality, it formerly having been moved every three years to the home country of the current General Secretary. As Edith Müller took over the post of General Secretary from George Contopoulos at the end of the Grenoble General Assembly, she handed to me the files concerning Symposia, Colloquia and Regional Scientific Meetings, the details of these important meetings being the domain of the Assistant General Secretary ; the files were in impeccable order and I had my car with me to take them back to Ireland.

At that stage I had not had occasion to concern myself closely with IAU

Symposia and Colloquia, although I remembered well the introduction around 1952 of scientific meetings marking the progress of astronomy and astrophysics into the formal activities of the Union, and the very successful and interesting Symposia included in the programme of the 1955 Dublin General Assembly. Especially uncertain to me in 1976 was the question of what powers were assigned to the Assistant General Secertary (AGS) in determining the conduct and content of meetings. Edith was able to show me very clearly how to interpret the standing instructions contained in *Rules for Scientific Meetings* so as to maintain a balance between individual initiatives from Commission Presidents, and others, and the criteria which the Executive Committee of the Union would expect to be met in each proposal put before them.

It was, and still is, the tradition of the IAU scientific meetings that the chief impetus comes from the "grass roots" and, in general, topics are not "dreamed up" by Executive Committee (EC) members meeting once a year. EC members do, however, cast very critical eyes upon the proposals put before them and it is the task of the AGS to ensure, as far as possible, sometimes on a tight time-schedule, that proposals are in good order before they come up for decision by the EC. The IAU Officers, at their meetings held two or three times a year, would review the proposals received and secure any further details that might be required. To maintain a balance between what proponents of a particular speciality desire and what the EC might think to be best suited to the interests of the Union was not always easy. Edith's talent for managing things and her suggestions to me as her successor were admirable in showing how practical ways for reaching agreement could be arrived at which, in the end, would cause few serious problems.

The Executive Committee met in Geneva in 1977 and were welcomed with gracious hospitality by l'Observatoire de Genève, as arranged by Edith. It happened that an invitation to the EC for 1978 was not then to hand. During one afternoon meeting I had to make an international telephone call to check a point under discussion and on my return was met by persuasive words from our President, Adriaan Blaauw, to the effect that it was thought that an invitation to Dublin for August 1978 would be very acceptable to the Committee. So it came about that I had to make arangements for a busy four-day meeting with participants *inter alia* from the USA, the Soviet Union, Poland, Australia, Greece and Canada, according particularly to Edith's requirements for herself as General Secretary and for the two IAU Secretariat Staff members. At the committee table, at the light luncheons, at the dinner reception, and on the "optional" trips (Birr Castle on Wednesday, Newgrange on the Saturday after the meeting), Edith was uniformly appreciative and cheerful and always ready to assist with any minor difficulty. She could be opinionated on particular topics at EC meetings, but always with sound opinions ; if there was another viewpoint she was always ready to listen.

A typical solar IAU meeting... At the right of Edith : Y. Chmielewski and R. de la Reza. At her left : J.C. Pecker.

In 1979 communication was mostly by letter, with occasional telephone calls and Telex, if appropriate. Fax and e-mail were then unknown; correspondence at the IAU Secretariat depended on deadline dates being met and decisions being made accordingly. When the time came for the Executive Committee to wind up its business at the end of a General Assembly, exhaustion of General Secretaries could be imminent and hasty decisions might go through. An example of this came at the end of the 1979 Montreal meeting when a decision was needed whether to support a proposal that had come in formally one day after the deadline for submission. This proposal involved close cooperation with another ICSU Union and was of a type for which there had been several precedents. Edith thought that the funding support needed could be usefully employed elsewhere and advised against acceptance. Having been unable to dissuade the Executive Committee otherwise, I was dismayed, as I took up the matter in the new Paris office, on reading the file in detail, to find myself convinced that the decision would have to be reversed. After consultation with the secretary of the other Union, I had a discussion with Edith on the telephone before mailing my recommendation to all other EC members. The reversal of the earlier decision, made when the "fatigue factor" had set in at Montreal, went through without undue dissent and the project turned out, over the forthcoming decade, to be eminently successful.

 It is a curious experience for a new IAU General Secretary to take up duty, having served as Assistant General Secretary for three years. From being decidedly the "junior" member of the EC, the new General Secretary becomes immediately, by an *instant* transformation, the person who has to be fully informed on every issue, from UNESCO politics down to any petty problems in the IAU Secretariat. For me the transformation came at the end of the Montreal General Assembly as Edith surrendered all her business and took up her role of Adviser to the EC. Already the new permanent secretariat staff had been chosen - Brigitte Manning as Secretary and Patricia Smiley as Assistant Secretary replaced the long-serving Arnost Jappel and his assistant, Jarka Dankova, both from Czechoslovakia. It had been arranged that I would join Edith in interviewing Brigitte in Paris in June 1979, but for some reason my flight out of London the previous evening had turned back and I could not reach l'Observatoire until the interview was over. (Edith seemed to think my non-attendance was due to

mismanagement on my part!) At that stage, in spite of Paris having been decided on for a full year, the actual arrangements for setting up the office did not seem to be well advanced. I met Brigitte over lunch and with trepidation we tried to assess how we might "get along" in the future - rather doubtfully, if I correctly recall. I need not have worried. Edith had chosen well and from "Day 1" in the quaint gate lodge that was to be the abode of the IAU Secretariat for quite a few years, there was never any discord between myself as the new General Secretary and Brigitte Manning as the new Secretary. With new staff, a new office environment, and a new President and Executive Committee, everything could be decided afresh if necessary. Edith as my principal Adviser was always ready sympathetically to advise, usually on the telephone, but my other official Adviser, Adriaan Blaauw, then at Leiden, was also readily available with excellent advice. Really, I was admirably supported, for as well as these two valuable colleagues, who were so immersed in IAU business, I could seek further advice from two other former General Secretaries with whom I had cordial personal relationship, Jean-Claude Pecker in Paris and Donald Sadler in Herstmonceux. I was exceedingly fortunate in my Advisers, official and unofficial. It became clear that IAU business in all its many ramifications, is a speciality *only* of former IAU General Secretaries (and some Presidents), and Edith Müller must have remained a lynch-pin as adviser for a considerable succession of General Secretaries, albeit in an unofficial capacity.

For me, both coming to a peak during my first year as General Secretary, the two biggest problems that I had to deal with were the finding of a venue for the 1982 General Assembly (having left Montreal with no certain country offering to come forward but with three possibilities - Spain, Greece, and the United States) and of finalising the renewed adherence of the People's Republic of China. Edith had spent years of her youth in Spain and was fluent in Spanish, as well as in English, French, German and *Schweitzerdeutsch*, and was sure that an invitation from that country could be secured. However, it was not to be - Greece filled the bill in the end, with a happy result for "Patras, 1982". Also, "The China Problem" had come to a head at Grenoble following a protest by several astronomers (especially by George Miley and Tao Kiang) in that the EC had not done enough to redress its "wrong" action in respect of admitting Taiwan to IAU membership in 1952. The quandary as to how the IAU could approach a country that had left the Union haunted the Executive Committee over the years 1976 to 1979. Edith had laid the foundation work that encouraged me to travel on behalf of the EC, with active suggestions from President Adriaan Blaauw, to Beijing in April 1979 to "suss out" the possibilities of agreeing a mode of renewed Chinese mambership without severing, as at least one Union had done, the formal ties with Taiwan. Recognising the merits of Taiwanese adherence, a modicum of agreement was reached in Montreal, particularly through the patience and skill of our President, Adriaan Blaauw, that required some more correspondence with and, in fact, concession on the part of, the *Chinese Astronomical Society* of Taipei, Taiwan, China. With added pressure from three other ICSU Unions, each following the formula that had been suggested in Montreal (jointly with the International Union of Biochemistry), agreement in principle was reached in 1980 in time for the 1980 EC meeting in Leiden. In due course the day came (in early 1981) when China paid over its annual contribution in Swiss Francs. Apart from formalities concluded in Patras, the solution had thus been worked out without precipitating a crisis situation. In both these traumatic episodes of 1979-1980, re-assurance and advice from Edith Müller was very important to me, although during her own time, while laying the foundations, she had been unable to conclude the negotiations, which took more time to produce the desired fruit.

At Armagh Observatory (1978). From left to right : M. de Groot, Mrs de Groot, C. Fehrenbach, A. Blaauw, Edith, G. Simms, Öpik, F. Byrne, D.H. Lilly, S. Grew; H.A. Brück, two students, J. Butler, P. Wayman at the occasion of a meeting of the E.C. of the IAU in Dublin. (Courtesy M. de Groot)

Over her long service to the IAU subsequent to 1982, including work for the Schools for Young Astronomers, the support of "new" IAU member countries, and administration of the IAU Exchange of Astronomers scheme, as well as Commission work, Edith never missed a General Assembly and continued to enthuse over many aspects in the tremendous developments that have affected astronomy since 1948, the year of the Zürich General Assembly. It was a particular pleasure to me and Mavis my wife to find ourselves in the same hotel as Edith at the 1994 General Assembly in The Hague and to be able to exchange so many memories over the breakfast table, where we found Edith's appreciation neither of international astronomy nor of a good Dutch breakfast had lessened with the years. For some time we had heard from her of the diminished capability of her sister and how on her retirement from Geneva she had moved to Basel, the city of her youth, the more easily to see after the needs of her ailing sister. It was apparent how much she dreaded any prospect that she herself would, for long tedious years, be dependent upon the ministrations of others. Therefore when, within a year of "The Hague", news came that Edith had died, active in the very task she excelled in - making highly personal contributions to a scientific meeting - we were glad that her end in this world had, finally, been as she would have wished even if so many tasks were left to be finished. In her long professional life she had done much to foster international good will and, although she was never a radical feminist, to be an example that would serve to strengthen the role of female humanity among the real contributors to our science. She was always highly critical of "false prophets" within our scientific communities, of either sex and of any political persuasion, but only with good cause would she condemn. It was my valued privilege to follow her in the service of the IAU, the international scientific body that especially epitomises before all others the "universality" of its science.

————— Derek McNALLY (University of London, England) —————

FROM ONE G.A. TO ANOTHER...

I first met Edith in 1970 at the Brighton General Assembly of the IAU. It was a meeting destined to have fateful consequences, for the IAU. Prior to 1970 I was able to enjoy General Assemblies with complete freedom of choice. I could attend scientific meetings as my interests dictated. Little did I know all that was to change; at the Brighton General Assembly, I became aware of Commission 46 The Teaching of Astronomy. I was a teacher of astronomy, so I went along. It was an interesting meeting academically but I said too much and I left as putative Vice-President for the next triennium; having been pressed hard jointly by Edith and Bart Bok, who could withstand such a pressure group?

Commission 46 was established at the first General Assembly I ever attended, Hamburg in 1964. The founding President was E.L. Schatzman. He had foreseen the need for the IAU to become involved in the education of astronomers and in what has now come to be called the Public Understanding of Science. Clearly the IAU could not enter into the organisation of national education, each country has its own individual approach, but it could provide a forum where educational matters could be discussed, ideas exchanged and help provided to raise international standards. Commission 46 first appeared at the Prague General Assembly in 1967. The Commission Meetings provided a forum whereby the fledgling Commission could define its purposes and develop a programme. Its Vice-President was then E.A. Müller. By the time of the Brighton General Assembly, Edith was its President. But by 1967 its effect was being seen in the form of a summer school for young astronomers, the first of what became known as International Schools for Young Astronomers. It was of 6 weeks' duration and subsequent plans were for 8 week long schools, times were less fraught in the late 60s and early 70s; in our more frenetic times a school is thought to be "long" at three weeks.

Evry and Edith between them developed a series of projects for Commission 46, Astronomical Educational Material, The Visiting Lecturers Project, Project Contratype (organised by Kharadze and developed by Gerbaldi and Kononovitch) as well as the ISYA (organised then by Kleczek) mentioned earlier. They also foresaw the need to associate Commission 46 with the Education Committee of ICSU and to foster relations with UNESCO.

So by the time I was attending Commission 46 meetings in 1970 in Brighton, there was a well defined programme already in existence. It seems that what ensnared me into the Vice-Presidency was my suggestion that there could be help available through disposal of duplicate books and journals and possibly of actual teaching equipment. So there I was at the end of the Brighton General Assembly enmeshed in the gears of the IAU.

It was to be a valuable and valued learning process for me. From Edith I learnt that IAU affairs did not happen, they had to be driven. We worked together to produce a framework for the Commission and Guidelines for the ISYA. We had to argue hard with several ECs that the ISYA were a good investment for the future in times which were just beginning to become harder for science, halcyon days though they were, as we now see with the benefit of hindsight. Edith was generous in her support for the ideas of others, I particularly remember her valued support when I proposed that Commission 46 should meet with local teachers at a General Assembly. We introduced the concept of a one day meeting for local teachers as part of the programme of the General Assembly in Sydney. The idea was not well received by the Executive Committee! IAU meetings are for IAU members and Invited Participants, not for the public. It looked like an impasse. But Edith steered us to a compromise. She persuaded the EC that if a meeting was held with teachers the day before or the day after the General Assembly, no problem would arise. So in Sydney we instituted the traditional meetings between Commission 46 and the local teachers that are now so much a part of the General Assemblies. These meetings take differing forms to meet with local circumstances but all fulfill our original aim to bring teachers and astronomers into contact. I learnt a valuable lesson in Sydney, it matters very little what the formal programme was about, what mattered was that local teachers got together and formed a group. Where isolation had been the norm, reinvention of the wheel was rampant, teachers got together to discuss problems, concepts, methods of interest and to prod professional astronomers into thinking seriously about how to present astronomical data in formats suitable for adaptation in school teaching. Teachers today who attempt to teach astronomy as a part of physical/mathematical science are much better supported than in 1970 and much more aware of the resources available to them. In this context access to the World Wide Web is an exciting revolution in access to resource material if not yet an optimum way to present such material in a coherent and structured teaching methodology. There is still a social function in the Teachers'Meeting to allow teachers to meet together and to meet the astronomical community. The second such meeting at the Grenoble GA in 1976 produced CLEA, one of the most successful manifestations of C46/Teacher interactions.

I was by then President of IAU 46. French is not my best "foreign" language (there are some who say the same applies to my "English") and how I longed to have Edith's skill with languages. Her mastery was simply and plainly daunting. I could remember how she had handled meetings in Sydney in at least four languages easing the flow and spontaneity of the discussion with superb almost simultaneous translation skill still keeping a grip of the meeting as Chairman. Her consummate skill in switching from one language to another impressed me with the importance of being able to communicate across language barriers. Even though English now seems to be the dominant language of international meetings, its many international dialects can still generate profound misundertandings, not least between the UK and USA ! Edith became General Secretary of the IAU in Grenoble. There had been awareness that the peripatetic nature of the IAU Secretariat was burdensome on a number of different but interacting fronts, not least among these was the inefficiency it gave to the administration of the IAU itself. Edith took up the challenge, and negotiated a home for the IAU Secretariat. France offered accommodation in a Gatehouse of the Observatoire de Paris. It was a location evocative of astronomical progress and social history of Astronomy and had a 17th century graciousness. The General Secretary from his/her office could view the forecourt and front of the Paris Observatory and, if time permitted (which was not often) could day dream just a little about the major figures of international astronomy who not only worked there but visited from abroad down the centuries. The Secretariat eventually outgrew its historic location and moved to more functional offices in the building of the Institut d'Astrophysique at the back of the Observatory. More efficient yes, but as the General Secretary who first occupied the new accommodation, I felt a certain loss of the graciousness of the original location. The IAU owes a great debt to Edith as General Secretary who negotiated the location for the permanent Secretariat and above all to the French Government, who in the end made it all possible.

Earlier, I said that General Assemblies were never to be the same after meeting Edith at that Commission 46 meeting in Brighton. Hereafter General Assemblies entailed *responsibilities*, as Vice-President, President, past President of Commission, and just when all that seemed over and I could enjoy the science and the sun in Patras, I found myself once again, at the Delhi General Assembly, enmeshed in the gears of the IAU, as Assistant General Secretary and General Secretary. Once again I became acutely aware of what my predecessors had done for the Union and its impact on astronomical science. One became aware of how, through all manner of political upheaval and international crises, the community of astronomers worldwide had held together. Divided by language, ideology, race, all the major alleged impediments to good human relations, it is a great encouragement to see all such inhibitions abandoned in the cause of astronomical cooperation and debate. Such international friendliness and good fellowship does not happen on its own. Edith had that happy knack of making everyone feel welcome, old, young, from any part of the world, and of drawing them into active participation. She, with many other likeminded astronomers, have done a superlative job in improving and preserving good international relations, finding solutions to seemingly the most intractable of political problems. Astronomy has an essence which transcends mere diplomacy!

I met Edith for the last time at the General Assembly in The Hague in 1994. She told me of her heart condition and how she had decided not to seek medical intervention. She told me how she had enjoyed her life to the full and was not prepared to accept a life of restriction, perhaps as an invalid. Characteristically she died, still enjoying life on holiday in a country she found highly congenial : Spain. I sadly had to admire her courageous decision, so completely in keeping with her outlook, though I am sure that all of us who were fortunate enough to be not just colleagues, but friends, might wish she had decided otherwise so that we could still benefit from her practicality and warm good sense.

———————— Alan H. BATTEN (Victoria, B.C., Canada) ————————
EDITH AND THE IAU : ONE BIG FAMILY.

Many of us who began our astronomical careers in the late 1950s or early 1960s first became aware of Edith Müller when the paper on abundances of elements in the Sun, that she wrote together with Leo Goldberg and Lawrence Aller, appeared. Of course, she already had an established reputation as a solar astronomer, but that paper certainly brought her to the attention of a wider group, especially amongst us younger ones, and probably ranks as one of her most important contributions. In due course, I came to know all three authors personally, but, as it happened, Edith was the last one I met. Since we were both assiduous in our attendance at IAU General Assemblies, our paths must have nearly crossed several times, but it was at the relatively small fifteenth General Assembly, held in 1973 in Sydney, Australia, when she was about to become Assistant General Secretary, that I first remember encountering her. I recall her at one commission meeting urging the president to submit a symposium proposal to the new Assistant General Secretary -and then stopping in midsentence as she remembered (or pretended to remember !) that she would be on the receiving end of this proposal.

 Edith's election to the IAU Executive coincided with my own increasing participation in the affairs of the Union, both as a Commission President and in the organization of the seventeenth General Assembly in Montreal in 1979. Since Edith was the General Secretary in the three years leading up to that Assembly, we were in frequent contact and spent some time together in Montréal, both immediately before the Assembly and about eighteen months in advance of it. Thereafter, we always got together at every subsequent Assembly, except the one in Buenos-Aires to which, for some reason, she was unable to come. Strangely, we never met except at, or in connection with, IAU General Assemblies.

Edith could sometimes look rather severe, even formidable, and one had to get to know her to appreciate her charm and humour. For me, the opportunity to do that came during the period of preparation for the Montréal General Assembly. When she declared that my strongly accented French was actually easier for her to understand than that of native-born Québecois, I felt that we were getting on famously. Any lingering doubts were dispelled that evening over an excellent meal in a French restaurant (I do mean "French" and not "Francophone" !) in Old

Through all Commission 46 meetings at Brighton, being new Commission members, we had been under the impression of Edith Müller's strong personality, and especially of her ability to grasp immediately what the new Commission members have been trying to express while speaking timidly in English, this being not their mother language. And Edith Müller herself has had a special facility of switching over from one language to another, from English to French, from French to German. Already from the beginning she had expressed much concern about the tasks of individual Commission members, National Representatives of many countries, among whom she distributed a handwritten xerox-copied letter : "As national representative in Commission 46 you are responsible and have the duty to inform all astronomical institutions and schools interested in astronomy of the existence, the work and the various projects of our Commission..."

Since 1970 I have met Edith A. Müller during many General Assemblies, where she had always some duties to perform, up to the task of General Secretary of the IAU. I am very grateful to note that at the General Assembly in Patras in 1982, acting with her usual warm friendliness, she had expressed much concern about the conditions in my country, Poland, then under martial law.

I remember especially vividly our meeting during the General Assembly in November 1985 in New Delhi. Edith Müller arrived there with her arm encased in plaster of Paris. She ought to remain home, in Switzerland, but she found it her duty to come for the ten days to India, the more so since many astronomers involved in university teaching had to be absent. We were sitting side by side during the Opening Ceremony held in the presence of Prime Minister Radjiv Gandhi. There had been extremely severe security requirements, we had to leave our conference bags as well as handbags before entering the spacious Hall. I think we were allowed to bring only the ceremony programmes and reading glasses. I tried to help Edith Müller with her belongings. And we were listening for two hours to the opening speeches, to the words of welcome, to the presentation of Indian poetry and music... I was specially thankful to Edith Müller when she came later to the business sessions of Commission 46 when I was acting President of the Commission, and she became extremely helpful in explaining to the audience the adopted membership rules and other business matters.

During 1985-1988 Edith Müller had been President of Commission 38 (Exchange of Astronomers), and I had written to her some letters related to various meetings planned for the General Assembly of 1988 at Baltimore. We were all sorry that owing to organisational work Edith Müller could not attend Colloquium N° 105 held at Williamstown before the General Assembly at Baltimore, since this had been the first Colloquium in the history of the IAU devoted to astronomy teaching.

It was at the session on the History of the IAU during the General Assembly in the Hague in 1994 that I met Edith Alicia Müller for the last time. Competent as always, she explained clearly some details of the past history of the IAU. Who would have supposed that she had only one year more to live ?

My contributed paper presented at the second IAU meeting devoted to astronomy teaching in July 1996 in London - IAU Colloquium N° 162 - was dedicated to the memory of Edith A. Müller. I brought participants her photograph for them to be acquainted with her engaging smile, a symbol of her personal charm and warm friendliness.

———————— John R. PERCY (Ontario, Canada) ————————

EDITH A. MÜLLER AND IAU COMMISSION 46

Introduction

Education is important to astronomy because it affects the recruitment and training of the next generation of astronomers, and it affects the awareness, understanding, and appreciation of astronomy by the taxpayers and politicians who support us. Astronomy is also important to education (above and beyond its contribution to the profession of astronomy) because of its practical and philosophical applications, its consequent role in history and culture, its contributions to mathematics and sciences, and its own status as one of the most dynamic of the sciences. In the classroom, it can be used to demonstrate the role of observation in the scientific method, and it can be used very effectively to teach many aspects of physics. It attracts young people to science and technology, and increases public interest in science - important considerations in both the industrialized and the developing countries.

The International Astronomical Union is the world organization of astronomers - a non-governmental union founded in 1922 "to promote and safeguard astronomy... and to develop it through international co-operation". The IAU had no formal body for dealing with astronomical education until the 1960's, when several astronomers -notably Edith Müller- worked to establish  a special commission or interest group dealing with education. IAU Commission 46 (The Teaching of Astronomy) is now one of 40 commissions, and is the only one which deals exclusively with education. Considering that the Commission has now existed for a third of a century, it is an excellent time to pay tribute to the role which Edith Müller played in its birth and growth.

The work of IAU Commission 46

The purpose of IAU Commission 46 is "to further the development and improvement of astronomy education at all levels, throughout the world". It co-operates with other scientific and educational organizations (such as the United Nations, the International Council of Scientific Unions, the International Planetarium Society) ; it works through its National Representatives in the countries which adhere to the IAU, to promote astronomy education in those countries ; and it encourages all other programs which support its purpose.

Commission 46 communicates with the astronomical community through its sessions at IAU General Assemblies, its regular reports in the IAU *Transactions*, its two colloquia (in 1988 and 1996) and their proceedings, its Triennial Reports, its Newsletter (both electronic and paper), and its home page on the World Wide Web (http ://physics.open.ac.uk/IAU46/). I encourage you to use these to find out more about the work of the Commission.

The Origin of Commission 46

By the 1960's, the IAU was discussing seriously the restructuring of the Union and its Commissions. The Editor of this volume, Jean-Claude Pecker, was a key figure in these discussions. The idea of a formal body within the IAU to deal with education is first mentioned in the report of the IAU Executive Committee for 1961-63 : "a discussion on the teaching of astronomy is being organized during the General Assembly in Hamburg ; among the administrative matters to be discussed will be whether the Union should set up a Commission (or a Committee) on the Teaching of Astronomy, and whether the Union should participate in the work of the Inter-Union Commission on Science Teaching" (Trans. IAU XIIB, 1964). The outcome of the discussion is reflected in the proceedings of the 1964 General Assembly :"... the Executive Committee formally proposed to the General Assembly that the following changes be made in the Commissions of the Union : (i) the creation of the two following new Commissions :... No. 46 : L'Enseignement de l'Astronomie (The Teaching of Astronomy)". Evry Schatzman was appointed as the first President of the Commission ; no Vice-President was appointed.

 It is interesting to note that the only reference to education in Adriaan Blaauw's *History of the IAU* is an account of the education session at the 1964 General Assembly in Hamburg :"Among the most important features of the Assembly was the attention given to teaching of astronomy in a meeting on August 29. The proceedings were published in full in *Transactions XIIB*, p. 629-649. The introductory paper by the chairman, M.G.J. Minnaert, included the items Astronomy in the Secondary Schools, The training of Future Astronomers at the University, Astronomy as a Minor Topic for Mathematicians and Physicists, Astronomy for Students of All Faculties, and Recruitment of Professional Astronomers. Discussions followed, introduced under the titles General Comments, International Co-operation in Astronomical Teaching, Practical Exercises in Astronomy, and Teaching of Radio Astronomy". These are topics which are still of interest today, though the work of Commission 46 has gradually focussed, more and more, on the needs of the developing countries.

The evolution of Commission 46

It is remarkable to witness the rapid development of the Commission, as expressed in its Triennial Reports, and in the reports of its meetings during the General Assemblies. Almost all of the present activities of the Commission came into being during those first few years. Edith Müller was deeply involved in that process.

119

Evry Schatzman served as President until partway through the meetings at the 1967 General Assembly. An Organizing Committee was formed. Edith Müller was appointed Vice-President, then almost immediately assumed the presidency. She continued as President through the 1973 General Assembly, then as an *ex-officio* member of the Organizing Committee in 1973-76. In 1976-79, she became General Secretary of the IAU (where she was undoubtedly able to keep the interests of Commission 46 in mind) ; she was a member of the Executive Committee from 1973 to 1982. This was one of the most crucial periods in the IAU's history, because it was the period during which, after intense and complicated negotiations, the People's Republic of China rejoined the Union. Edith Müller also served as President of Commission 38 (The Exchange of Astronomers) in 1985-88. This Commission works closely with Commission 46, and with the IAU Working Group on the Worldwide Development of Astronomy, to provide opportunities for young astronomers, especially from the developing countries, to work abroad.

Subsequent presidents of Commission 46 have been : Derek McNally (1973-76), Edward V. Kononovich (1976-79), Donat G. Wentzel (1979-82), Leo N. Houziaux (1982-85), Cecylia Iwaniszewska (1985-88), Aage Sandqvist (1988-91), Lucienne Gouguenheim (1991-94), and John R. Percy (1994-97).

By 1970, the Commission had established its present system of National Representatives. One of their responsibilities is to prepare Triennial Reports on the state of astronomy education in their countries. These reports make useful and interesting reading. They are sent to the National Representatives for local distribution, and also posted on the Commission's Web page.

Also by 1970, the Commission began a project to compile and distribute a triennial list of Astronomy Educational Material (AEM). This large undertaking, which was initially edited by Edith Müller, was continued for two decades. By that time, listings of English-language material were being compiled by others (such as the Astronomical Society of the Pacific's *Universe at your Fingertips*), and listings of material in many other languages were being produced nationally. There is still a need for listings in French and especially Spanish, for the benefit of the many countries (other than France and Spain) which use these languages.

The Commission also developed *Project Contratype*, which collected useful educational material (books, slides, photographs, etc.) and lent them to those who did not have access to such material themselves. The cost and unreliability of mail service (and the reluctance of some borrowers to return the material) made this project difficult, and it was terminated some years ago.

In 1973, at the General Assembly in Sydney, Australia, Commission 46 organized a one-day meeting between astronomers and 70 local school teachers. Such meetings have become an important and successful tradition : almost 200 teachers attended the one-day meeting at the 1994 General Assembly in the Netherlands. Teachers' meetings have been held at several IAU Regional Meetings, and even at an IAU scientific colloquium, on stellar pulsation, in South Africa.

The "flagship" project of the Commission is the International Schools for Young Astronomers (ISYA). These are intensive three-week schools attracting 20 to 50 students, mostly at the gradute level. They are held approximately once a year, in different parts of the developing world. About half the students normally come from the host country, the rest from elsewhere. A handful of faculty members give mini-courses on topics chosen by the host institution, in consultation with the ISYA secretary. Students often maintain contact with the instructors (and with each other), long after the ISYA is over, so the Schools are very effective in providing a network of contact people for the students.

The ISYA began at about the same time as the Commission ; the first three were held in Manchester (1967), Arcetri (1968), and Hyderabad (1969). For two decades, the Secretary was Josip Kleczek of Czechoslovakia. Donat Wentzel is presently Secretary.

The Commission also discussed the need for a program of visiting lecturers to countries in which astronomy was taking hold. These ideas led to a formal Visiting Lecturers Program (VLP), which was instrumental in developing astronomy in Paraguay and Peru in the last decade. The VLP has since been replaced by a more flexible program called Teaching for Astronomical Development (TAD), which is presently active in central America, and Vietnam.

Commission 46 Today

Almost all of the present activities of Commission 46 can be traced back to the six years in which Edith Müller was President. One new and successful initiative, however, has been the two colloquia on astronomy education (in Williamstown USA in 1988, and in London UK in 1996); another colloquium is being planned for 1999. Another recent initiative, carried out following a suggestion by Derek McNally and Richard West, was a "Travelling Telescope" to provide simple research and training facilities for countries which lack them. The logistics of transporting the telescope has proven to be a problem, but the instruments have been used on small local telescopes (presently in Paraguay), with some success. The Commission is also now cooperating with the UN which, with the European Space Agency, has been organizing a series of Workshops on Basic Space Science, in the various parts of the developing world.

The Internet and the World Wide Web have proven to be efficient and effective for communication, but the Commission is painfully aware that, for many countries, they are either unavailable, or prohibitively expensive. Much of the Commission's work still depends on the "old" methods of communication -mail, and personal visits.

The contributions of Edith Müller

Although Edith Müller was not the founding President of Commission 46, she was President for the six crucial years in which the Commission took root, and grew. The formal reports on the founding of the Commission are brief and neutral, but strong persuasion was certainly necessary to convince the Executive Committee to establish and fund a variety of new initiatives related to education and development. In the words of Derek McNally (1997, private communication), "Edith... had a very clear concept of what the IAU could do, and she fought off a considerable number of Executive Committee challenges, especially on the International Schools for Young Astronomers in those early years". As pointed out by Edward Kononovich (1997, private communication), she was able to gather the right people around her, and encourage them to work effectively. She maintained a good balance between being gently persuasive, and being strong and decisive in situations requiring immediate action.

The rules and guidelines of Commission 46 were also developed during Edith Müller's term as President; they were written primarily by her and Derek McNally, during the 1973 General Assembly. Drafting such documents, and ensuring that they are approved, is a tedious and difficult process, but absolutely necessary for the long-term survival of any organization. Edith Müller's organizational skills, and her fluency in many languages, contributed to her success both as President of Commission 46 and as IAU General Secretary. Commission 46 is technically a Subcommittee of the IAU Executive Committee, and its main projects (ISYA, VLP, and TAD) are funded by special grants from the Executive Committee. Therefore Edith Müller's service on the Executive Committee in 1973-82 was an important continuation of her years of service to the Commission.

Epilogue

It is fair to say that, without Edith Müller's time, effort, and thought, the birth and growth of IAU Commission 46 would have been much slower. No one played a more significant role in its evolution than she did (though several individuals come close - Derek McNally and Donat Wentzel are two names which come first to mind). Her enthusiasm and wisdom enabled the Commission to grow quickly, and establish its main projects securely.

Still, there is much to be done. Only one IAU Commission out of 40 deals exclusively with education. Only two IAU conferences out of several hundred have been on that topic. There are still dozens of countries around the world which are in need of astronomical development. I urge every reader to consider how they could assist in astronomy education, and the work of Commission 46 - thereby continuing the legacy which Edith Müller left behind.

Acknowledgements. I thank Adriaan Blaauw, Edward Kononovich, and Derek McNally for their helpful comments.

──────── Evry SCHATZMAN (Meudon & Paris, France) ────────
ASTRONOMY IN EDUCATION, AND... SWISS WATCHES.

In 1971, the New York Academy of Science and the American Astronomical Society organized a meeting on the History of Modern Astronomy, in memory of Professor Marcel Minnaert. On that occasion, Edith Müller recalled that Professor Minnaert was deeply convinced that education and the teaching of science, and especially the teaching of his science, were fundamental to human progress. She adopted that conviction, and it is also in this spirit that I have myself approached the importance of Astronomy in Modern Education.

Of course, I have met Edith many times before and after this meeting. But memory is failing. It seems to me that I met Edith Müller for the first time a long time ago. Was it beginning of the sixties, end of the fifties ? I do not remember ! I began to visit Geneva Observatory regularly around 1965, and I have always been greeted by Edith. In 1965, as I wanted to buy a Swiss watch, she was kind enough to take me to one of the most famous makers in Geneva. Several times, I was invited by Edith to have dinner at her home and could appreciate her excellent cooking ! In 1976, being a representative of the French Physical Society, I attended the European meeting in Helsinki, and found Edith acting efficiently as a General Secretary for the European Physical Society.

The last time I saw her was at the meeting in honour of Golay, who was retiring. During the dinner, it happened that we were sitting together. Pleasant evening, pleasant memory ! I did not see her anymore after she retired. Going to Geneva, going to the Observatory, I missed the possibility of seeing her. She has been for me one of the highlights of Swiss astronomy.

——————————— Edward KONONOVITCH (Moscow, Russia) ———————————
BRIEF ENCOUNTERS, HERE AND THERE

I was still a student, but I knew quite well that the group at the McMath-Hulbert Observatory, and at the University of Ann Arbor in Michigan, were active on the study of the chemical composition of the Sun. As an outcome of this large programme, Goldberg, Müller and Aller had published their famous paper in the Astrophysical Journal Supplement in 1960. I had imagined, because of some inaccuracies in the translations into Russian of foreign publications, which were more numerous at that time than they are now, that the three authors were indeed their particularly eminent men of sciences. And I was very pleasantly surprised when, at the occasion (at Prague, in 1967) of a commission meeting, I found myself sitting beside the lovely Edith Müller.

My knowledge of English was still quite poor ; and I could not follow the thoughts of the speaker without some moments during which I was at a loss...I turned towards my neighbour ; and I was deeply impressed to see how deeply she was listening, completely absorbed in her careful attention to the talk. No force could have taken her away from it....! There was, in her attitude, not only an active understanding, but a deep respect for the speaker...

I then already knew that Evry Schatzman, with the strong help of Marcel Minnaert, had constituted, at the preceeding General Assembly of the IAU, the new Commission 46, devoted to the Teaching of Astronomy. When I was told that the following President of that commission would be Edith Müller, I did not have any trouble to select the meetings I should absolutely not miss.

The Organizing Committee of the Commision 46 met in a comfortable small restaurant, on the place Stro, at Prague. And one understood immediately that the newly elected President was in the same time a brillant speaker and a clever organizer. She understood quickly the motivation and the substance of any proposal, she gave to it a concrete form, the project became alive, and took immediately its due importance and its full meaning.

Later, we have exchanged much correspondance (the e-mail did not exist yet !). These working letters were clear, to the point, and very detailed. To build, from scattered data, a coherent picture of the status of the rather meager teaching of astronomy in the world, to help, with our very modest means, that kind of teaching, in as much as we could do it, within our strengths, - that was much work, but a really fascinating one.

Later again, we met twice. The first one, at the General Assembly of Montréal, for the organization of which Professor Edith Müller, then General Secretary of the IAU, had played an major role. She gave to the Russian delegation a particular attention. When I entered her secretarial office, she asked me : *"Do you know who is the most famous astronomer in your country ?"*... and without waiting for a reply, she said : *"Victor Abalakin. He speaks so many languages !"*... She, too, spoke many languages. One knew her everywhere. One loved her everywhere...

Our last meeting took place in Moscow. I took her to the airport of Cheremetievo. We queued, waiting for her departure. And in the queue, we spoke about Pushkin. She regretted not to be able to read it in Russian ; she was eager to learn Russian through Pushkin, and she asked me to send her a volume of his poetry. Needless to say, I've tried to do my best.

————————— K. SINHA (Lucknow, India) —————————
MY "SCIENCE MOTHER"

I met Professor Müller in 1985 during the IAU Symposium on Astrochemistry in Goa (India). I found Professor Müller extremely encouraging, inspiring and sympathetic towards younger colleagues. I had the privilege of discussing a paper of mine with her during the symposium, though it was not very interesting nor a great paper from me. Still, I had a patient hearing and guidance from her.

During lunch, we talked about several things including Indian concept and ceremony for marriage. She was particularly impressed by the thought that the Hindus, who believe in rebirth, consider the matrimonial relationship not as *"till death do us part"* but for seven births. She was so overwhelmed that while mentioning that she never married, she felt very happy to see so many of *"my science children"*. Since then I always addressed her in my correspondence - I could never meet her again - as my *"science mother"* and she always responded with *"my Indian Science son"*.

I was completely taken aback and felt shocked to know of her sad and sudden demise. I distinctly remember her affection showered on me when she brought a gift of a plate full of desserts with one of her arms in plaster. It is very very difficult to express the gratefulness to her. To record my association with late Professor Müller, it was possible for me to dedicate one of my papers to her memory during a sun-related workshop held at Udaipur (India) in October, 1996.

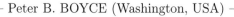

Peter B. BOYCE (Washington, USA)
DO I REMEMBER EDITH MÜLLER ?

When the editors of this volume asked me if I remembered Edith Müller, the answer was, *"Of course"*. Our paths intertwined several times.

Yes, I remember Edith. I remember her contributions to the American Astronomical Society's Chrétien Awards program on the award Selection Committee, serving as chairperson for several years. In Edith's usual fashion, that meant taking on much of the work and doing it fairly and thoroughly.

Yes, I remember Edith, at IAU meetings, as an old friend, but also as Secretary General, shouldering the burden of organizing the General Assembly, and running everything in her usual well-organized fashion.

But, most of all I remember Edith at the University of Michigan during my time as a graduate student - from 1959 to 1963. Edith was renowned among the graduate students, first as a friend and a fellow evening worker at the observatory. Many is the time we would all troop out for a bite and a cup of coffee during an evening break. And Edith was always a good friend to each of us.

She was also famous for her Fasching parties to celebrate the beginning of

Lent. Each February the faculty and students would invade Edith's little house in wondrous costumes to eat, drink and have a good time. Everyone made their own costumes then. There was no such thing as running out to a costume store and renting a lookalike mask and fancy duds which instantly turn you into some currently fashionable character. No, sir! I don't think Edith would have allowed that. She organized the contest in which the goal was to be the last person recognized and unmasked.

We all scoured our attics and trunks for days ahead of time, fashioning our own costumes, and trying desperately to be the last person recognized. Couples would park two blocks away and arrive at Edith's front door by separate routes and at separate times to further disguise their identity. And, through it all, Edith would keep the party going, maintain a filled puch bowl and, best of all for us poor graduate students, keep a seemingly never-ending supply of goodies coming. She was a wonderful hostess and contributed greatly to making the Michigan Astronomy Department feel more like a family than any other place I have worked.

It seems funny now, but in 1960 we regarded Edith as one of the "different" solar astronomers. They were the people who used to work in the basement of the old Observatory measuring "wiggly lines;" that is when they weren't up at the Lake Angeles solar observatory taking incredibly large spectra. Edith was our prime contact with the solar community; you know, those astronomers with the peculiar habit of observing during the daytime.

But it was when Edith gave a course in solar physics that she really shone. It was a marvelous course. Edith was well organized, thorough and she presented the material with insightful clarity. She helped us students to understand what the "wiggly lines" were telling us about the Doppler motions on the sun's surface. Through her presentation we could see first-hand what limb darkening did to the continuum and spectral line energy distribution. It made us aware for the first time the enormous simplifications we were using to analyze the spectra from the nighttime stars. And it was breathtaking to trace the course of solar flares and prominences using real films from the Michigan H-alpha patrol.

I think this was the one course which brought together all our astrophysics knowledge and cemented our understanding of the stars. And yet, Edith Müller,

128

made it seem logical, easy and fun. She was an outstanding teacher, yet I think she only gave the course once. All the students were grateful for the circumstances which put us in her course.

But, I cannot reminisce about Edith without recalling her car. A Fiat Bianchina from the late 1950s, and one of the smallest cars I, or anyone else at the Observatory had ever seen. I think Leo Goldberg's lawn mower had a bigger engine.

There was a driveway behind the Observatory, and those of us "evening regulars" used to park in the driveway, but we always saved the prime space for Edith -which really didn't matter, since Edith's car was so small. One time, someone blocked Edith's car in. I clearly remember when she said 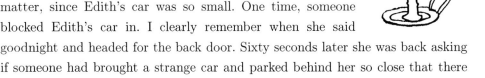 goodnight and headed for the back door. Sixty seconds later she was back asking if someone had brought a strange car and parked behind her so close that there was no hope of her getting out.

Well, no. The car did not belong to any of us. We all went out to see what could be done. Sure enough, she couldn't get out. What to do? That was the night six graduate students discovered that we could pick up Edith's car. We just applied our muscle power and moved that little car right out of its tight spot and sent Edith on her way. That was also the night that we looked at the Observatory's loading dock and realized that Edith's car would fit up on the dock sideways, with no room to spare.

 What a joke that would be for Edith to come out some night and find her car marooned up on the loading dock, with no apparent way to get it down. We laughed as we contemplated the sight of a car up there, but the students all agreed that we loved and respected Edith too much ever to play such a prank on her. I believe we still do...

Observatory and her satisfaction with the information she got there and which she approved.

Besides, I was under the impression of her keen interest in the history of the people of the country she visited for the first time, in its religion, ancient culture and arts, in national songs and general ethnology. During our dialogue, I couldn't note any sign of tiresomeness or any shadow of indifference.

As for her personal character we all remember firmly her warmth and charm, her friendliness and her unchangable readiness to share with you sympathy and kindness, being always ready to help especially young astronomers.

I do not say here anything about her distinguished hospitality, although I have had not only once the privilege of enjoying it in her beautiful and eminent country of Switzerland. I say nothing now also on what is very well known to the community of astronomers and to a wider scientific society ; that is, Professor Edith Müller's contributions to knowledge of the solar physical state and chemical composition and general solar activity.

En route from Georgia to Armenia, August 1981, in a valley of abandoned cave dwellings. Underlined (l. to r.) : A. Blaauw, E. Kharadze, R. Wilson, M. Wayman, E. Müller, P. Wayman (General Secretary), X, B. Manning (Secretary), T. West, M. Feast, V. Bappu (President), P. Smiley (Assistant Secretary), R. West (Assistant General Secretary). (courtesy P. Wayman)

——————— Brigitte MANNING (Caussade, Tarn-et-Garonne, France) ———————
DRINKING TEA IN GEORGIA

Edith was responsible for me entering the IAU as Executive Secretary. She was a wonderful person, generous, undestanding.. But can I tell the following story ?

 Once in Georgia, in 1981, we were taken from Tbilissi to the Abastumani Observatory, in the Caucasus, then directed by Kharadze, Vice-President of the IAU.... After many many hours, we paused in the country. It was deserted...There was not even a tree behind which to hide, in order to pay due tribute to nature.... Edith told me that this was one of her bigger problem when travelling. She was an addicted tea drinker. And she told me then that once, in Indonesia, where she attended a scientific meeting, they had been for hours and hours in a bus, - without a stop. When finally the busses stopped at the right place, the place in question was closed, - it was a Sunday !...

At a fine picnic in the Caucasus near Abastumani, August 1981. (courtesy P. Wayman)

——————— Yaroslav YATSKIV (Kiev, Ukraine) ———————
FROM OUR FIRST MEETING SHE CONFIDED IN ME.

My first meeting with Edith Müller.

During the years 1973-1976, Professor Edith Müller was Assistant General Secretary of the IAU, and responsible, in that capacity, for the organization of Symposia and Colloquia, as well as for the Regional Meetings of the IAU. As there are always, at the IAU, more proposals for meetings than can actually be adopted, it is necessary for the Assistant General Secretary to be particularly scrupulous, and strictly objective. As a rule, one presents proposals for organizing IAU symposia in due time, i.e. at least two years ahead of time. However, during the XVI-th General Assembly of the IAU (at Grenoble, France, 1976), Professor Walter Fricke and myself were forced, because of various circumstances, to ask Edith Müller to make an exception for our project, and to accept that Symposium n^o78, *"Nutation and the Earth's rotation"* be held in Kiev, as early as in May 1977.

Edith Müller met us in her office, listened very carefully, and made clear to us that very little time was left to prepare the Symposium. She listened again to our argument : I gave her the assurance that the Local Organizing Committee would do its utmost for the success of the endeavour. She then gave her agreement for the holding of this Symposium, - and she convinced the Executive Committee.

This first meeting with Edith Müller gave me a very profound impression. Firstly, without being a specialist in the field of astrometry and the theory of the rotation of the Earth, she displayed a deep understanding of the problems involved. Moreover, while I was for her a young  and unknown scientist, she relied upon me and she thought that I would indeed be able to organize in such a short time an IAU Symposium in Kiev....

The challenge.

The need to change the system of astronomical constants appeared to be connected with the need to compute high accuracy ephemerides of the Moon and planets, and with the introduction of a new fundamental system of coordinates, such as achieved by the catalogue FK5. That new system of astronomical constants had been prepared by the ad hoc working group constituted to that effect, under the chairmanship of W. Fricke, and had been discussed during the IAU Joint Discussion between Commissions 4, 8, 31, at the XVIth General Assembly of the IAU. The main result of these debates had been expressed in Resolution n^o1, by which the IAU approved the new system of astronomical constants (IAU, 1976). In particular, it was specified, by Recommendation n^o4, that *"the tabular nutation shall include the forced periodic terms listed by Woolard for the axis of figure in place of those given for the instantaneous axis of rotation"*.

This recommendation was not voted by the unanimity of present Members, and started a very animated discussion on two points :

(i) In any high accuracy computation of the nutation, it is indispensable to take into account the influence of the elasticity of the mantle, and of the existence of a fluid core in the Earth ; this meant it was absolutely necessary to elaborate a new theory for the nutation, in order to replace Woolards theory.

(ii) One must re-examine the question of the choice of the axis for which the equations of nutation must describe the motion.

At Kiev, Professor E. Fedorov and his pupils were actively busy with these questions. This is precisely why I had asked to hold in Kiev a special IAU Symposium devoted to the problem of nutation. This proposal became particularily to the point, and urgent, because of the creation of the FK5 ; and this is why the idea was actively supported by W. Fricke.

The Kiev Symposium became a great success. It contributed to solve a very complex scientific problem, and to adopt a new theory of nutation (IAU, 1980).

Concluding.

This minor episode in the life of Edith Müller could appear as quite insignificant. But I still remember the open kindness, and the intelligence of the way this remarkable woman looked at the people in front of her, of her quiet manners, of her capacity to listen to others...

———————— G. SWARUP (Pune, India) ————————
EDITH AND THE GIANT RADIOTELESCOPE

I first came across Edith in 1979 in Montreal. She inspired me to pursue my proposal for the Giant Equatorial Radio Telescope. Unfortunately it did not come through, but she was very helpful in encouraging me to pursue the matter for several years. She was also instrumental in assisting the very successful IAU meeting in 1985 in New Delhi.

———————— Peter WILSON (Sydney, Australia) ————————
FROM "DOWN UNDER"

I had the privilege of meeting Edith Müller at IAU and other scientific meetings during the 1960s and 70s. I remember her as a charming and gracious lady whose contributions to these meetings were always acute and helpful. Unfortunately, I was not directly associated with her in any scientific projects but I remember, with gratitude, her friendship and her encouragement to a young scientist from the antipodes.

———————— Alla G. MASSEVICH (Moscow, Russia) ————————
IN GENEVA, ONE DAY...

In the early 80s, I met Edith Müller in Geneva. I was there at a meeting of the World Peace Committee and I used this opportunity to call Edith. She was at that time already retired, but she took me to the Observatory and acted as a beautiful guide. She also was very helpful with sightseeing and shopping in Geneva. We had a very nice time together.

Edith was a very active, life-asserting, joyous and many sided personality, a prominent scientist and a very charming women.

Back in Andalusia : a joyful dinner (1974). (courtesy E. Alfaro)

——————————— Yoshihide KOZAI (Tokyo, Japan) ———————————

EDITH A. MÜLLER MISSED IN JAPAN

With many Japanese colleagues I very much missed Professor Edith A. Müller. Indeed many of our friends here remember well her remarkable scientific contributions, particularly on accurate determination of solar abundances. In early days after the World War II when Japanese astronomers started to go abroad to study newly developed astronomical research made during the war, a few of them stayed at Ann Arbor (Michigan) and I heard from them about her achievements. In fact in those days whenever anybody came back to Japan from abroad all the staff and students attended meetings where they reported about what they learned abroad. Those days it was a dream for us Japanese, to go abroad and it was only possible when one could find any funds to do that, usually from abroad. Therefore, although I have never been a specialist in the fields where she had been working, I could recognize her name for many years.

She had been famous not only as an excellent astronomer but also as an administrator, as she was successful General Secretary of the IAU. In fact the

General Secretary is the most important and difficult position in the IAU. I happened to be the Japanese national representative and the president of a commission while she was the General Secretary. For me it was a very impressive event at a meeting of commission presidents during the General Assembly when she opposed a paper which condemned the absence of several important scientists of east-european countries. The paper was submitted to the General Secretary by several presidents of solar-system related commissions and it was distributed among us. I believe that she understood much better the situation in east european countries and after that there was no further argument about the paper. Indeed she was a very able General Secretary of the IAU and I had admired her very much for her contributions to astronomy.

Edith Müller.

Edith A. Müller (1918-1995) – List of publications

Müller E.A., "Gruppentheoretische und Strukturanalytische Untersuchungender Maurischen Ornamente aus der Alhambra in Granada", Ph. D. doctoralthesis, University of Zürich, 1943. Buchdruckerei Blaublatt A.G.,Rüschlikon (Zürich), 1964.

Goldberg L., Müller E.A., "The vertical distribution of nitrous oxide and methane in the Earth's atmosphere", J. Optical Soc. America **43**, 11, 1033-1036, 1953.

Goldberg L., Müller E.A., "Carbon monoxide in the sun", ApJ **118**, 3, 397-411, 1953.

Goldberg L., Dodson H.W., Müller E.A., "The width of H-alpha in solar flares", ApJ **120**, 1, 83-93, 1954.

Goldberg L., Aller L.H., Müller E.A., "The abundances of the elements in the sun", in "Stellar Atmospheres", Proc. Indiana Conference on Stellar Atmospheres, 141-146, M.H. Wrubel ed., Indiana Univ.,1954.

Müller E.A., "Die Verdoppelung der Dimensionen im Weltall, ORION, Mitt. Schweiz. Astronom. Ges. **42**, 198-201, 1954.

Becker W., Müller E.A., Steinlin U., "Dreifarben-Photometrie des Offenen Sternhaufens NGC 7510 und seiner Umgebung", Zeitschr. f. Astrophys. **38**, 81-94, 1955.

Müller E.A., "Dreifarben-Photometrie des Offenen Sternhaufens NGC 654 und einer benachbarten Sterngruppe", Zeitschr. f. Astrophys. **38**, 110-124, 1955.

Goldberg L., Müller E.A., Aller, L.H., "The chemical composition of the solar atmosphere", AJ **62**, 15-16, 1957.

Müller E.A., "Die erste visuelle Beobachtung der Satellitenrakete (Sputnik 1) in Ann Arbor, Michigan, am 13. Oktober 1957", ORION, Mitt. Schweiz. Astronom. Ges. **58**, 325-327, 1957.

Goldberg L., Mohler O.C., Müller E.A., "The profile of H-alpha during the limb flare of February 10, 1956", ApJ **127**, 2, 302-307, 1958.

Goldberg L., Mohler O.C., Müller E.A., "The double reversal in the cores of the Frauenhofer H and K lines", ApJ **129**, 1, 119-133, 1959.

Goldberg L., Müller E.A., Aller L.H., "The abundances of the elements in the solar atmosphere", ApJS **45**, 5, 1-137, 1960.

Müller E.A., Mutschlecner J.P., "Center-limb effects on solar abundances", AJ **67**, 5, 277-278, 1962.

Müller E.A., Mutschlecner J.P., "Effects of deviations from local thermodynamic equilibrium on solar abundances", ApJS **9**, 85, 1-64, 1964.

Müller E.A., "A critical discussion of the abundance results in the solar atmosphere", IAU Symposium 26 on "Abundance Determinations in Stellar Spectra", 171-199, H. Hubenet ed., Academic Press London, New York, 1966.

Müller E.A., "The composition of the solar atmosphere", in "Solar Physics", 33-67, J. Xanthakis ed., John Wiley and Sons Interscience publishers, London, New York, Sydney, 1967.

Müller E.A., Baschek B., Holweger H., "Center-to-limb analysis of the solar oxygen lines", Solar Physics **3**, 125-145, 1968.

Müller E.A., "Lithium observations in the sun", in Highlights of Astronomy, 243-246, L. Perek ed., D. Reidel Publ. Co. Dordrecht, Holland, 1968.

Müller E.A., "The solar abundances" in "The origin and distribution of the elements", 155-176, L.H. Ahrens ed., Pergamon Press, Oxford, 1968.

Müller E.A., "Ballonexperimente zur Erforschung des infraroten Sonnenspektrums", Actes Soc. Helv. Sci. Nat. **148**, 66-68, 1968.

Schatzman E.L., Müller E.A., Page T.L., "Report of meetings of Commission 46 on the Teaching of Astronomy", Transactions IAU XIIIB, 223-229, 1968.

Stettler P., Kneubühl F., Müller E.A., "Ballon-Interferometer zur Messung der Sonnenstrahlung mit Wellenlangen zwischen 10 μ und 1mm", Helv. Phys. Acta **42**, 630-631, 1969.

De la Reza R., Müller E.A., "Sur la formation des raies de Fraunhofer en dehors de l'équilibre thermodynamique", Actes Soc. Helv. Sci. Nat. **150**, 127-129, 1970.

Chmielewski Y., Müller E.A., "Décomposition numérique d'un blend et application au doublet de résonance de BeII dans le spectre solaire", Actes Soc. Helv. Sci. Nat. **150**, 129-131, 1970.

Müller E.A., "Neues über die Lithium und Beryllium Haufigkeit in der Sonnenatmosphare", Actes Soc. Helv. Sci. Nat. **150**, 132-133, 1970.

Müller E.A., "Astronomy Educational Material", I.A.U. Commission 46 on the Teaching of Astronomy, Special Publication, Observatoire de Geneve, vol I, 1-172, 1970.

Müller E.A., "The development and the present state of astronomy education in different countries", I.A.U. Commission 46 on the Teaching of Astronomy, Special Publication, Observatoire de Geneve, vol. II, 1-58, 1970.

Müller E.A., "The teaching of astronomy, Report of Commission 46", Transactions IAU, XIVA, 559-566, Dordrecht, Holland, 1970.

Chmielewski Y., Müller E.A., "Sur le problème des opacités continues solaires dans le proche ultraviolet", Actes Soc. Helv. Sci. Nat. **151**, 41-42, 1971.

De la Reza R., Müller E.A., "Formation de la raie de résonance du Potassium dans la photosphère solaire", Actes Soc. Helv. Sci. Nat. **151**, 43-44, 1971.

Müller E.A., Swihart, T.L., "Reports of meetings of Commission 46 on the Teaching of Astronomy", Transactions IAU, XIVB, 236-241, Dordrecht, Holland, 1971.

Stettler, P., Kneubühl F.K., Müller E.A., "Absolute measurement of the solar brightness in the spectral region between 100 and 500 microns", A&A **20**, 309-312, 1972.

Müller E.A., "Professor Marcel Gilles Jozef Minnaert (1893-1970)", in "Education in and History of Modern Astronomy", R. Berendzen ed., Annales of the New York Academy of Sciences **198**, 5-7, 1972.

Müller E.A., "The Commission on the Teaching of Astronomy of the International Astronomical Union", in "Education in and History of Modern Astronomy", R. Berendzen ed., Annals of the New York Academy of Sciences **198**, 66-76, 1972.

Holweger H., Müller E. A., "The photospheric Barium spectrum : solar abundance and collision broadening of BaII lines by hydrogen", Solar Physics **39**, 19–30, 1974.

Stettler P., Rast J., Kneubühl F.K., Müller E.A., "Far–infrared solar brightness measured with a balloon–borne lamellar–grating interferometer", Solar Physics **40**, No 2, 337–349, 1975.

Müller E.A., Brault J.W., "The solar Lithium abundance. I. Observations of the solar Lithium feature at 6707.8Å", Solar Physics **41**, No 1, 43–52, 1975.

Müller E.A., Peytremann E., De la Reza R., "The solar Lithium abundance. II. Synthetic analysis of the solar Lithium feature at 6707.8Å.", Solar Physics **41**, No 1, 53–65, 1975.

Müller E.A., Stettler P., Rast J., Kneubühl F.K., Huguenin D., "The solar brightness temperature in the far infrared", Osservazioni e memorie dell'Osservatorio di Arcetri (First European Solar Meeting), Firenze **105**, 3–5 1975.

Chmielewski Y., Müller E.A., Brault J.W., "The solar Beryllium abundance", A&A **42**, 37–46, 1975.

De la Reza R., Müller E.A., "The Potassium abundance in the solar photosphere", Solar Physics **43**, No 1, 15–32, 1975.

Müller E.A., Jappel A., Editors of : Transactions of the International Astronomical Union, Grenoble, Reidel Dordrecht, 586 pages, 1977.

Müller E.A., Editor of : Highlights of Astronomy, vol. 4, parts I and II, Reidel Dordrecht, 370 pages and 307 pages, 1977.

Lamers H.–J., Müller E.A., Llorente de Andres F., "Line blocking in the near ultraviolet spectrum of early–type stars. I. Observed line blocking factors for 132 stars", A&AS **32**, 1–16, 1978.

Rast J., Kneubühl F., Müller E.A., "Measurement of the solar brightness temperature near its minimum with a balloon–borne lamellar–grating interferometer", A&A **68**, 229–238, 1978.

Müller E.A., "Reports on Astronomy", in : Transactions of the IAU, vol. XVII, E.A. Müller ed., Reidel Dordrecht, 1979.

Barambon C., Müller E.A., "On asymmetries of solar spectral lines", Solar Physics **64**, 201–212, 1979.

Kneubühl F., Rast J., Stettler P., Müller E.A., Huguenin D., "Minimum far infrared solar brightness temperature and sunspot activity", in : 4th Intl. Conf. on "Infrared and Millimeter Waves and their Application", S. Perkowitz, Miami Beach Florida, 177–178, 1979.

Rast J., Cartier F., Kneubühl F., Huguenin D., Müller E.A., "Measurements of the absolute solar brightness temperature in the far–infrared with a balloon–borne interferometer", A&A **83**, 199–200, 1980.

Müller E.A., Kneubühl F., Rast J., Stettler P., "Variability of the far–infrared solar temperature minimum with the solar cycle", A&A **87**, L3–L4, 1980.

Castelli F., Lamers H.J.G.L.M., Llorente de Andres F., Müller E.A., "A comparison between the observed and predicted UV line–blocking for blanketed model atmospheres of early–type stars", A&A **91**, 32–35, 1980.

Cartier F., Kneubühl F., Huguenin D., Müller E.A., "Submillimetre–wave brightness temperatures of solar active zones and sunspots measured from a balloon–borne platform", ESA SP189, 175–179, 1982.

NAME INDEX